Ric

The Inter

Cinderella

Springer
*Berlin
Heidelberg
New York
Barcelona
Hong Kong
London
Milan
Paris
Singapore
Tokyo*

Jürgen Richter-Gebert
Ulrich H. Kortenkamp

The Interactive Geometry Software
Cinderella
Version 1.2

With 126 Figures and a CD-ROM

Springer

Jürgen Richter-Gebert
Institut für Theoretische Informatik
ETH Zentrum
CH-8092 Zürich
email: richter@cinderella.de

Ulrich H. Kortenkamp
Institut für Informatik
Freie Universität Berlin
Takustraße 9
14195 Berlin, Deutschland
e-mail:
kortenkamp@inf.fu-berlin.de
kortenkamp@cinderella.de

Die Deutsche Bibliothek – CIP-Einheitsaufnahme
The *interactive geometry software Cinderella* [Medienkombination]: interactive geometry on computers / J. Richter-Gebert; Ulrich H. Kortenkamp. – Berlin; Heidelberg; New York; Barcelona; Hong Kong; London; Milan; Paris; Singapore; Tokyo: Springer.
ISBN 3-540-14719-5   Buch. 1999   CD-ROM. 1999

Mathematics Subject Classification (1991): 51-04

ISBN 3-540-14719-5 Springer-Verlag Berlin Heidelberg New York

This work is subject to copyright. All rights are reserved, whether the whole or part of the material is concerned, specifically the rights of translation, reprinting, reuse of illustrations, recitation, broadcasting, reproduction on microfilm or in any other way, and storage in data banks. Duplication of this publication or parts thereof is permitted only under the provisions of the German Copyright Law of September 9, 1965, in its current version, and permission for use must always be obtained from Springer-Verlag. Violations are liable for prosecution under the German Copyright Law.

© Springer-Verlag Berlin Heidelberg 1999
Printed in Germany

The use of registered names, trademarks, etc. in this publication does not imply, even in the absence of a specific statement, that such names are exempt from the relevant protective laws and regulations and therefore free for general use.

Please note: All rights pertaining to the Software (program and handbook) are owned exclusively by Springer-Verlag. The software is protected by copyright. The publisher and the authors accept no legal responsibility for any damage caused by improper use of the instructions and programs contained in this book and the CD-ROM. Although the software has been tested with extreme care, errors in the software cannot be excluded. Decompiling, disassembling, reverse engineering or in any way changing the program is expressly forbidden. For more details concerning the conditions of use and warranty we refer to the *License Agreement* in this book (pp. 131ff.).

Legal note: Java and all Java-based trademarks and logos are trademarks or registered trademarks of Sun Microsystems, Inc. in the United States and other countries. Springer-Verlag and the authors are independent of Sun. For details concerning the Java Runtime Environment license see Sect. 8.3, pp. 139ff. All other brand and product names are trademarks, registered trademarks or service marks of their respective holders.

Awards: Multimedia-Transfer '97: Most innovative Multimedia software
Finalist of the European Academic Software Award (EASA) 1998

Microsoft and the Microsoft Internet Explorer logo are registered trademarks or trademarks of Microsoft Corporation in the United States and/or other countries.

Photocomposed from the authors' HTML file.

Cover design: *design & production*, Heidelberg

SPIN 10784753   46/3143 - 5 4 3 2 1   Printed on acid-free paper

Contents

1 *Preface* 1
2 *Introduction* 3
   2.1 Sample Applications 4
      2.1.1 Exact Drawings 4
      2.1.2 Geometric Calculator 5
      2.1.3 Student Exercises 5
   2.2 Design and Features 6
   2.3 Technical Background 9
3 *A Quick Start* 11
   3.1 Pappos' Theorem 12
      3.1.1 Drawing Your First Point 12
      3.1.2 Undoing an Operation 13
      3.1.3 Moving a Point 13
      3.1.4 Adding a Line 14
      3.1.5 Adding More Lines 14
      3.1.6 Creating Points of Intersection 15
      3.1.7 Finishing the Drawing 16
      3.1.8 Selecting and Changing the Appearance 16
      3.1.9 Adding a Final Point and Proving a Theorem . . . 18
      3.1.10 Moving Points to Infinity 19
   3.2 A Three Bar Linkage 22
      3.2.1 Making a Bar 22
      3.2.2 Adding Two More Bars 23
      3.2.3 Moving the Construction 25
      3.2.4 Starting an Animation 26
      3.2.5 Drawing a Locus 26

vi   Contents

4 *Behind the Scenes* 28
   4.1 Problems in Interactive Geometry 28
      4.1.1 Static Problems 28
      4.1.2 Dynamic Problems 29
   4.2 Projective Geometry 31
   4.3 Homogeneous Coordinates 32
   4.4 Complex Numbers 34
   4.5 Measurements and Complex Numbers 37
      4.5.1 Euclidean and Non-euclidean Geometry 37
      4.5.2 Cayley-Klein Geometries 38
   4.6 The Principle of Continuity 41

5 *Reference* 44
   5.1 Overview 44
      5.1.1 The Menu 45
      5.1.2 The General Action Toolbar 45
      5.1.3 The Geometric Tools 46
      5.1.4 The Geometries 47
      5.1.5 The Views 47
   5.2 General Tools 48
      5.2.1 File Operations 48
         5.2.1.1 New 48
         5.2.1.2 Load 48
         5.2.1.3 Save 48
         5.2.1.4 Save As 48
      5.2.2 Export Tools 48
         5.2.2.1 Print All Views 48
         5.2.2.2 Export to HTML 48
         5.2.2.3 Create an Exercise 49
      5.2.3 Undo/Redo 49

5.2.3.1	Undo	49
5.2.3.2	Redo	49
5.2.3.3	Delete	49
5.2.4	Selection Tools	49
5.2.4.1	Select All	49
5.2.4.2	Select Points	50
5.2.4.3	Select Lines	50
5.2.4.4	Select Conics	50
5.2.4.5	Deselect	50
5.2.5	Moving an Element	51
5.2.6	Select	53
5.2.7	Interactive Modes	54
5.2.7.1	Add a Point	55
5.2.7.2	Add a Line	57
5.2.7.3	Line Through Point	59
5.2.7.4	Add a Parallel	60
5.2.7.5	Add a Perpendicular	62
5.2.7.6	Add a Line With Fixed Angle	63
5.2.7.7	Add a Circle	64
5.2.7.8	Circle by Radius	65
5.2.7.9	Circle by Fixed Radius	66
5.2.7.10	Midpoint	67
5.2.8	Definition Modes	68
5.2.8.1	Center	69
5.2.8.2	Angular Bisector	70
5.2.8.3	Compass	71
5.2.8.4	Mirror	72
5.2.8.5	Circle by Three Points	73

5.2.8.6 Conic by Five Points	73
5.2.8.7 Polar of a Point	74
5.2.8.8 Polar of a Line	74
5.2.8.9 Polygon	75
5.2.8.10 Join	76
5.2.8.11 Meet	76
5.2.8.12 Define a Parallel	77
5.2.8.13 Define a Perpendicular	78
5.2.9 Measurements	79
5.2.9.1 Distance	80
5.2.9.2 Angle	81
5.2.9.3 Area	82
5.2.10 Special Modes	83
5.2.10.1 Add Text	83
5.2.10.2 Locus	85
5.2.10.3 Animation	87
5.2.10.4 Add a Segment	89
5.3 Geometries	91
5.3.1 Types of Geometries	91
5.3.2 Views and Geometries	93
5.4 The Views	94
5.4.1 Euclidean View	94
5.4.1.1 Translate	94
5.4.1.2 Zoom in	95
5.4.1.3 Zoom out	95
5.4.1.4 View All Points	95
5.4.1.5 Toggle Grid	95
5.4.1.6 Toggle Axes	95

5.4.1.7	Toggle Snap	96
5.4.1.8	Denser Grid	96
5.4.1.9	Coarser Grid	96
5.4.2	Spherical View	96
5.4.2.1	Rotate	98
5.4.2.2	Spherical Reset	98
5.4.2.3	The Scale Slider	98
5.4.3	Hyperbolic View	98
5.4.4	Polar Euclidean and Spherical View	98
5.4.5	Construction Text	100
5.4.6	General Functions	101
5.4.6.1	Generate PostScript	101
5.4.6.2	Euc Hyp Ell Choose the Geometry	102
5.5	The Appearance Editor	103
5.5.1	Color	104
5.5.2	View Colors	105
5.5.3	Clipping	106
5.5.4	Labelling	107
5.5.5	Pinning	107
5.5.6	Overhang	108
5.5.7	Size	108
5.5.8	Opaqueness	109
6	***Creating Interactive Webpages and Exercises***	110
6.1	Glossary	110
6.2	Exporting Plain Examples	111
6.3	Exporting Animations	112
6.4	Creating Interactive Exercises	113
6.4.1	Exercise Construction	113
6.4.2	Editing the Exercise	114

x Contents

      6.4.2.1 Defining the Input 114
      6.4.2.2 Defining Solutions 116
      6.4.2.3 Defining Hints 118
      6.4.2.4 Tool Selection 119
   6.4.3 Saving and Creating the HTML 119
   6.4.4 Testing the Exercise 120
   6.4.5 Design Considerations 120
 6.5 Post-Processing 121
 6.6 Legal Issues 122

7 Installation 123
 7.1 General Information 123
 7.2 Installing on Windows 95, 98 or NT 4.0 123
 7.3 Installation on Unix Platforms 123
   7.3.1 Installing on Sun Solaris (SPARC) 123
   7.3.2 Other Unix-like Platforms 124
   7.3.3 Installing a JVM on Linux 124
 7.4 Installing on MacOS 124
 7.5 Installing on Other Java Platforms 125
 7.6 Installing Using a Web Browser 125
 7.7 Troubleshooting 125

8 License Agreement 126
 8.1 Conditions of Use and Terms of Warranty 126
 8.2 Nutzungs- und Garantiebedingungen 130
 8.3 Java(tm) Runtime Environment 134

9 References 137

1 Preface

Cinderella is a program for doing geometry on a computer. In its present form it is the product of a sequel of three projects done between 1993 and 1998. It is based on various mathematical theories ranging from the great discoveries of the geometers in the nineteenth century to newly developed methods that find their first applications in this program.

The idea for the first of these projects was born in 1992 during a combinatorics conference at the Mittag-Leffler Institute in Sweden, when Henry Crapo and Jürgen Richter-Gebert were taking a trip on a boat called "Cinderella." At that time Jürgen Richter-Gebert had developed symbolic methods for automatic theorem proving in geometry [RG], and both of them dreamed of a software where one could input geometric configurations with just a few mouse clicks and then ask the computer about properties of these configurations.

Henry and Jürgen started the project on a NeXT platform which at that time was famous for its marvelous software architecture. ***Cinderella*** became the working title for the project and this title turned out to be unremovable from the project.

A few weeks of development produced the first working prototype. The program was based on principles from projective geometry and invariant theory. It was able to find *readable* algebraic proofs for many theorems of projective geometry about points, lines and conics [CrRG].

However, as a platform NeXT gradually declined in popularity, and with it the initial enthusiasm for ***Cinderella***. After the summer of 1995 almost no further progress was made. At a conference on computational geometry in Mt. Holyoke, MA, it was almost impossible to give a software demonstration due to the vanishing of NeXT computers and their operating system.

In August 1996, right after that Mt. Holyoke conference, we (Ulli Kortenkamp and Jürgen Richter-Gebert, at that time working at the Technical University of Berlin in the group of Günter M. Ziegler) decided to start a new project, based entirely on the platform independent language Java^{TM}. At that time the language Java was relatively new, and at first both of us were very sceptical about using an interpreted (presumably slow) language as the basis for a program that requires lots of computation in real-time. But we tried anyway.

The goal of this second project was to have the old functionality (that was available in the NeXT version) substantially extended by features of Euclidean and non-euclidean geometry. We also wanted functionality for geometric loci. Moreover, since Java is designed to be "internet-aware," the new program should be able to run inside a web browser. In particular, we wanted to be able to create student exercises for the web. The theorem proving facilities of the program should be used to automatically check the correctness of the student's solution.

Conferences, competitions and their deadlines are often driving forces for rapid development. A first working version was presented at the "CGAL-startup-meeting" in September 1996 at the ETH Zürich. A second version won the "Multimedia Innovation Award" at the Multimedia Transfer of the ASK Karlsruhe in January 1997.

During 1997 Jürgen Richter-Gebert became an assistant professor at the ETH Zürich. This change forced another break in the development. Ulli Kortenkamp moved to Zürich in September.

At the same time we began negotiating about the publication of *Cinderella*. Originally we planned to polish and finish the second project. However, things turned out differently.

The second version, like other computer programs for geometry, suffered from seemingly unavoidable mathematical inconsistencies. These inconsistencies came from ambiguities in operations like "Take the intersection of a circle and a line." There may be two, one or no intersections depending on the position of the circle and the line. While dragging a construction the program has to decide which of them to chose. This seemingly innocuous ambiguity may lead to terrible inconsistencies in the behavior of a construction. It may happen that while you move a point only a little bit whole parts of the construction flip over.

At the beginning of 1998 it turned out that this problem of jumping elements was indeed solvable. However, it was clear that implementing the theory would not be an easy job. Every configuration had to be embedded in a complex vector space. Results of analytic function theory had to be used to avoid "singular situations." If we wanted to use those new insights we had to rewrite the mathematical kernel of the program from scratch. The program had to perform approximately 20-100 times more computations, a challenge for us and for Java.

We decided to do this and ended up with the third project whose outcome you see here. In a period of unbelievably intensive work (that stretched our patience and that of our families to the extreme) we rewrote the whole program again - thereby tuning the program to higher performance at every possible place.

It turned out to be a good idea to undertake this effort. The benefits of the newly developed theory were much greater than we had originally thought. Based on the new methods we were able to do reliable randomized theorem checking. This proved to be much more useful than the old symbolic methods. It was also possible to generate complete geometric loci by generic methods, which is a novelty to the best of our knowledge.

The present program is a mixture of old geometry from the nineteenth century, complex analysis, our new methods and modern software technology. We hope you will enjoy it as much as we do.

Jürgen Richter-Gebert & Ulli Kortenkamp
Zürich, December 1998

2 Introduction

Why did we write *Cinderella*? Aren't there enough programs that are suitable for doing mathematics, and in particular for producing mathematical graphics? Well, there are many programs, but *Cinderella* is special in many respects.

We want to point out the major features of this software. *Cinderella* ...

- *... is a mouse-driven interactive geometry program:* With a few mouse clicks you can construct simple or complicated geometric configuration. No programming or keyboard input is necessary. After you have completed a construction you can pick a base element with the mouse and drag it around, while the entire construction follows your moves consistently. This enables you to explore the dynamic behavior of a drawing.
- *... has built-in automatic proving facilities:* While you construct your configuration *Cinderella* reports any non-trivial facts that occur.
- *... allows simultaneous manipulation and construction in different views:* You can view and manipulate the same configuration in the usual Euclidean plane, on a sphere and even in Poincaré's hyperbolic disc.
- *... has "native support" for non-euclidean geometries:* In *Cinderella* you can easily switch between Euclidean, hyperbolic and elliptic geometry. Depending on the context, your actions are always interpreted correctly.
- *... has advanced facilities for geometric loci:* The unique mathematical methods of *Cinderella* guarantee that complete real branches of the loci and not only parts of them are drawn.
- *... is "internet-aware":* The entire program is written in Java. Each construction can be exported immediately to an interactive web page. Even student exercises and animations can be created that way.
- *... produces high-quality printouts:* You can generate camera-ready PostScript files of your drawings. This vector output is superior to screen-shot-like pictures and uses the full resolution of the printer.
- *...is based on mathematical theory:* The whole implementation has a mathematical foundation. The theories of the great geometers of the nineteenth century, as well as many new insights, make *Cinderella* a highly reliable and consistent tool for geometry.

2.1 Sample Applications

The application areas of *Cinderella* reach from pure Euclidean (and non-euclidean) geometry, via physics (optics, for example) to computational kinematics and CAD. The following sample applications present potential scenarios which you can benefit by using *Cinderella*.

2.1.1 Exact Drawings

Assume you are writing a scientific publication for which you need one or more figures. When the drawings are a bit involved it is usually almost impossible to make a perfect one on the first attempt.

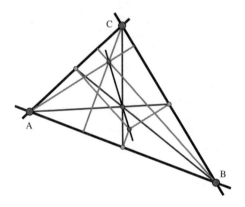

The Euler Line of a triangle

Either you will have many elements in almost the same place, or some parts of the figure will not fit onto the page.

With *Cinderella* you start by making a computer sketch of a construction. This sketch might not look like what you expect the final drawing, but it provides all relations that are decisive for the construction. Later you pick the base elements and move them around until the picture satisfies all your esthetic requirements. During the "move phase" you will always have a valid instance of your geometric construction. Finally you use *Cinderella's* Appearance Editor to adjust the color and size of each geometric element.

2.1.2 Geometric Calculator

Sometimes it happens that you want to get a feeling for some geometric situation. Either you have read something interesting in a geometry book or you have a new idea yourself.

You do the construction with *Cinderella* and start to play with it. Through geometric exploration you gain new insights and often hidden properties are revealed. *Cinderella's* mathematically consistent implementation ensures that no strange effects occur that do not come genuinely from the configuration.

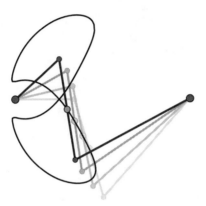

Dynamics of a three bar linkage

When you want to communicate your research to other colleagues you create an interactive web page and make it available on the internet. Then your colleagues have instant access to the configuration and can interact with it locally in their Java-enabled web browser.

2.1.3 Student Exercises

Another interesting application is the generation of interactive student exercises. Imagine that you want to teach students how to construct the circumcenter of a triangle using only ruler and compass. First you do the construction yourself. Then you create an interactive exercise by marking the "input elements", providing exercise texts, marking "intermediate construction steps" and the "final result". *Cinderella* generates an interactive Web page that presents the input elements (perhaps the triangle from which the students should start) together with all the construction tools for doing ruler and compass constructions.

The students can solve the exercises on their own computer and arrive at a solution completely by themselves or by following hints you provided. No matter which construction a student used to solve the problem, *Cinderella's* integrated automatic theorem checking facilities can decide whether it is correct or not. The student's creativity for finding a solution is not restricted by the program.

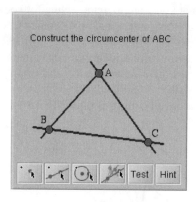

An exercise for students

2.2 Design and Features

Several major design goals lead us in the development of ***Cinderella***. We want to mention the three most important ones to give you an impression of the overall architecture of the program.

- *General approaches:* ***Cinderella*** is designed to cover a broad range of geometric disciplines. The program provides "native support" for *Euclidean Geometry*, *hyperbolic geometry*, *elliptic geometry* and *projective geometry*.

    This means that you do not have to simulate hyperbolic geometry by making complicated Euclidean constructions. You can use the "hyperbolic mode" of ***Cinderella*** and the constructions will behave like elements of the hyperbolic plane.

    Cinderella achieves this by implementing very general mathematical approaches to geometry that form a common background for all of the above areas. Much of the mathematics behind ***Cinderella*** makes use of the great, and unfortunately almost forgotten, geometric achievements of the geometers of the nineteenth century. To mention a few of them: *Monge* and *Poncelet* who "invented" projective geometry; *Plücker*, *Grassmann*, *Cayley* and *Möbius* who developed a beautiful algebraic language to deal with projective geometry; *Gauss*, *Bolyai* and *Lobachevsky* who "discovered" what is today called hyperbolic geometry; and finally *Klein*, *Cayley* and *Poincaré* who managed to get a unified description of Euclidean and non-euclidean geometries in terms of projective geometry and complex numbers. For an excellently written and exciting introduction to the historical development of geometry in the nineteenth century we recommend the book of Yaglom

[Yag]. Also the historical book [Kl1] written by Felix Klein himself is a very interesting introduction into this topic.

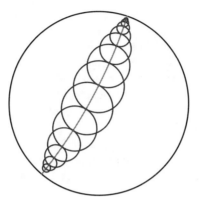

Hyperbolic circles of equal size

Projective geometry forms the background for the incidence geometry part of **Cinderella** and *Cayley-Klein geometries* form the backbone for the metric part of **Cinderella**.

- *Mathematical consistency:* To say it in a metaphor: "The geometric constructions done with **Cinderella** should behave as if they lived in a reasonable geometric universe. In this universe, no unnatural things should happen".

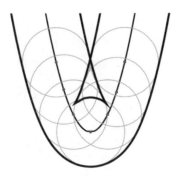

*Offset curve of a parabola -
a challenge for most CAD systems*

Other systems for interactive geometry suffer from mathematical inconsistencies. You do a construction, drag around the base elements and suddenly one part of the construction jumps from one place to another. This is unfortunately a usual scenario even in software systems for parametric CAD.

Cinderella completely resolves this problem by using a new theory. It makes use of features from *complex analysis* and merges them with the "old geometry" mentioned above.

Based on this theory it was possible to equip **Cinderella** with an *automatic theorem checker* that governs most of the internal decisions **Cinderella** makes. This theorem checker is also used for automatic feedback operations in student exercises. Another benefit of this approach is that you have a generic tool to construct correct and complete geometric loci, which are real branches of algebraic curves.

- *Modular design:* **Cinderella** is designed to be as modular as possible. This architecture makes **Cinderella** well prepared for further extensions in many directions.

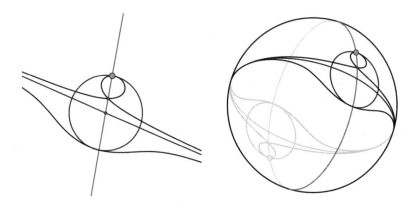

Conchoidal curve in a Euclidean *and in a spherical view.*

Even the present release benefits from the modular approach. For instance, it is possible to view the same geometric constructions simultaneously in many geometric contexts. A construction in hyperbolic geometry can be simultaneously shown, and manipulated, in the "Poincaré Model" of hyperbolic geometry and in the "Beltrami-Klein Model". The simultaneous use of different views helps to gain a deeper understanding of a configuration. For example, the "behavior at infinity" of a configuration becomes immediately visible in a spherical view.

2.3 Technical Background

Normally you would, and should, not care about the programming language and other technical details of software you just want to use. Nevertheless, we want to tell you about the computer science background of *Cinderella*.

Cinderella was written in Java, the platform independent programming language developed by Sun Microsystems. This means that the software can be run on any computer, irrespective of its operating system, provided that there is a thing called the "Java Virtual Machine" (JVM) for this system. These Java Virtual Machines are available from Sun Microsystems for Windows 95/98/NT and Solaris, and there exist ports for Linux, OS/2, MacOS, BeOS, AIX, HP-UX and many more. In fact, you probably already have a JVM installed on your computer, since Netscape Navigator and Microsoft Internet Explorer have a built-in JVM. This, in turn, means that you can run Java programs inside a web browser. These programs are known as "Applets".

We do not want to explain Java in full detail here, instead we recommend the official Java home page at http://www.javasoft.com as a starting point for further reading. However, we do want to tell you about some of the consequences of choosing Java for *Cinderella*.

It is a fact that, although Microsoft Windows is the dominating operating system today, many mathematics departments have a variety of Unix workstations. Even within the same working group you can find a mixture of different systems. Java enables everybody, regardless of the choice of platform, to use *Cinderella* in the same way. It is even possible to install and use the *same code on all of your computers*. For us, we could use our favorite operating system (Linux) for development, and at the same time we were sure to reach the largest possible audience.

Second, the fact that you can run Java software inside web browsers has been used for the web export functionality of *Cinderella*. This means that you can publish constructions easily, spice up your personal home page with animations, or assign construction homework to your students. Our license agreement [p. 126] gives you great freedom in redistributing the necessary parts of *Cinderella*, but please obey the few restrictions that come with it.

Java is an *interpreted* language, as opposed to *compiled* languages like C or C++, which are the usual languages used for most software. Interpreted languages have some technical advantages, but they suffer from an additional translation step which slows the program down. Java (or the virtual machine) has been tuned a lot for performance, and the performance gap is not as large as it was when our project started. Still we had to do a lot of optimization by hand to create acceptable speed, and the "interactive feeling", of *Cinderella*.

Sometimes you'll notice a short delay when you move a point. Do not blame your computer, Java or *Cinderella*. These delays are caused by extremely complex calculations which are necessary to get the correct result or the correct screen representation after a movement. The generation of correct loci is one reason for that; many intersections involving conics are the other. We tried our best to speed up these calculations, but there is a (mathematical) limit where we do not want to sacrifice accuracy for speed.

Finally we want to mention the tools that helped us in creating **Cinderella** and this documentation. First there is *XEmacs*, a powerful, extensible text editor, which is based on GNU Emacs, which in turn is a version of the original Emacs written by Richard Stallman in the seventies at MIT. It is definitely the best editor available, and we used it to write the whole program and all of the documentation.

The program itself was developed with the help of the *Java Development Kit* of Javasoft, a division of Sun Microsystems, in particular with the Linux port of it (see http://java.blackdown.org for more information on the Java-Linux porting project). *Linux* is a free, unix-like operating system originating from the work of Linus Torvalds, and is now continually improved by the effort of several hundred developers all around the world.

The parsing engine (used for loading saved constructions) was constructed with the help of *ANTLR 2.5.0*, a public-domain Java/C++ parser generator, written by Terence Parr of the MageLang institute.

Post-optimization and compression of the code was done with *Jax* from alphaworks, the research division of IBM. We want to thank the Jax team, in particular Frank Tip, for their help and IBM for the permission to use Jax commercially.

The *Concurrent Versions System (CVS)* by Cyclic Software did most of the version merging (and saved us from a lot of headaches). It is free software, too.

Thanks to the "browser war" between Microsoft and Netscape, the licensing terms for redistributing *Netscape Navigator* allow us to ship a Java-1.1 compatible browser with **Cinderella**.

The documentation of **Cinderella**, both the printed manual and the online version, were written with XEmacs in HTML. We used the same files for the printed version and the screen representation. The design of the web pages uses Cascading Style Sheets (CSS); the hardcopy was created using a whacked version of *html2ps* by Jan Karrman.

The icons and images used in **Cinderella** were designed by ourselves with *The GIMP* (GNU Image Manipulation Program), written by Peter Mattis and Spencer Kimball. In our view it is one of the most impressive freely distributed pieces of software. The additional figures in the documentation were created with **Cinderella**, of course, and *Povray*, a free 3D raytracing software, and some PostScript hacking.

Two people deserve special mention: *James Gosling*, the creator of the Java programming language, and *Jamie Zawinski*, responsible for the first Unix versions of Netscape Navigator. They are both connected to XEmacs in a special way: James Gosling did the first C-implementation of Emacs, known as GOSMACS, and Jamie Zawinski was the person responsible for XEmacs versions 19.0 to 19.10, which was at that time a collaborative work of Lucid (now out of business) and Sun Microsystems (sic!).

3 A Quick Start

The following two sections will guide you through large parts of ***Cinderella's*** functions. After working through the tutorials you should be able to use ***Cinderella***. You can look up other functions in the reference part [p. 44] of this manual. Most of the functions described there follow patterns that are explained in the tutorials.

Here are a few general instructions on how to use these tutorials.

- Read the texts carefully. When you come to a new section of the tutorial read it first completely. Then you should read it again, while you *perform the described actions*.
- Follow the instructions exactly. The tutorials ask you to do constructions in a certain order. If you do not stick to the same order then the labelling of your elements will be different from that in the tutorial.
- If you made a mistake there is always the chance to undo the last step by pressing the "Undo" button ⬅ [p. 49] in the tool bar. You can undo as many steps as you want.
- The pictures in the tutorials are intentionally screenshots, and not high-quality printouts. It should be easy to match these drawings with the actual situation on your screen.

Each tutorial focuses on a certain topic. The information provided by the tutorial text becomes less and less detailed. We assume that it is helpful in the very beginning to be told *precisely* what to do. The more advanced you get the more freedom you will have to do the operations.

- *Pappos' Theorem [p. 12]*
  This tutorial introduces you to the basic construction principles of ***Cinderella***. You will learn how to do a simple construction.
- *Three Bar Linkage [p. 22]*
  Here you will learn how to use the animation features of ***Cinderella***. You will also see how to produce geometric loci.

3.1 Pappos' Theorem

Pappos' Theorem is one of the most fundamental theorems in projective geometry. In a certain sense it is the smallest example of an incidence theorem. In this step-by-step example we will construct a draggable instance of the theorem.

3.1.1 Drawing Your First Point

When you start **Cinderella** the first window that shows up is a "Euclidean view." This window has a large tool bar that is the key to most of **Cinderella's** functionality. Below this toolbar is the drawing surface on which you perform the operations you want by constructing and dragging the elements you need.

You may notice that in the tool bar one button 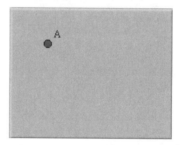 is slightly darker. This indicates the current mode of **Cinderella**.

All mouse action in the drawing surface refers to the selected mode. The mode you are in is the "Add a point" mode. *Move the mouse over the drawing surface and click the left button.* A new point is added and labeled with a capital letter.

Fig. 1: Your first point

Before you add a second point continue reading. If you hold the left mouse button down while adding a point then you can still move it around. The definition of the point will be adapted to the geometric situation at the current location of the mouse pointer (so far there is not a lot of geometry in our drawing but this will change soon). *Move the mouse over the drawing surface, press the left button and hold it. Now move the mouse.* You will notice that the new point sticks to the mouse pointer. Coordinates that tell the current position are shown. When you approach an already existing point (*try it*) the new point snaps to the old point. Only after you *release the button* the new point is added to the construction. If you release the mouse over an old point, no new point is added. *Play with these features and add a few new points.*

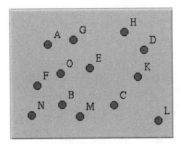

Fig. 2: Many points

3.1.2 Undoing an Operation

Your screen may look a bit crowded now. There is an undo-operation that inverts the actions you have performed. By pressing the undo button [p. 49] the last point you added will disappear. *Click the undo button until exactly two points remain on the drawing surface.* You can use the undo button whenever you make a mistake. You can undo as many consecutive operations as you want.

3.1.3 Moving a Point

We want to continue with our construction of Pappos' Theorem. Place the two remaining points *A* and *B* approximately at the position shown in Figure 3.

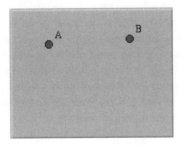

Fig. 3: Move to this situation

Most probably your points will not be there. To change this, *select the "move" mode [p. 51] by pressing the button in the tool bar.* Now, mouse actions in the drawing surface no longer add points; instead, you can pick points and move them around. *Move the mouse pointer over the point you want to move. Press the left mouse button. Hold it and drag the mouse.* The point follows the mouse pointer. You will also notice that in this mode it does not snap to points. When you *release the mouse button* the point is placed again. In general, the move mode allows free elements of a construction to be moved. The rest of the construction will change accordingly.

3.1.4 Adding a Line

We are now going to add a line from ***A*** to ***B***. *Switch to the "Add a line" mode [p. 57] by pressing the button* 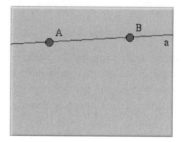 . In this mode you can add a line between two points using a press-drag-release sequence with the mouse. *Move over point **A**. Press the left mouse button. Hold it and move the mouse over point **B**. Release the mouse.*

Fig. 4: Add a line

This action should have created the desired line. You might have noticed a few things. When you pressed the mouse over point ***A*** the point became highlighted. This means that you selected this point as the starting point of the line. While you drag the mouse a second point is always present at the mouse position. When you release the mouse button somewhere this second point will be added. When you approach another point it will be highlighted and the second point snaps to it. You used the second action for attaching the line to ***B***.

If you did not notice these things or made a mistake, undo your operations and try again. Your final picture should look like Figure 4 before you proceed.

3.1.5 Adding More Lines

Now you are ready to add three more lines to end up with Figure 5. You can do this with just three mouse operations. *First, move over point **B** and press-drag to the position of the not yet existing point **C**. Do another press-drag-release from **C** to **D**, and finally from **D** to **E**.*

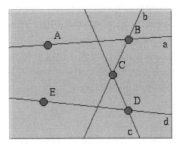

Fig. 5: More lines

Observe that you have done all this without leaving the "Add a line" mode. You did some unnecessary work when you added the points *A* and *B* in the "Add a point" mode. You could have done this directly in the "Add a line" mode, since not only will the second point be created if necessary, but also the first one. The lines created so far were automatically labelled with the lower-case letters *a* to *d*.

3.1.6 Creating Points of Intersection

We stay in the "Add a line" mode. Now we want to draw a line from point *E* to the intersection of the lines *a* and *c*. *Move the mouse over E. Press the button. Hold it and move to the intersection of a and c. Release the mouse.*

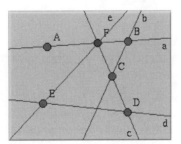

Fig. 6: Adding an intersection

When you approached the intersection of the lines they were highlighted and the endpoint of the new line snapped to the intersection. When you released the mouse the new point was defined to be the intersection of the lines. When you will move points *A* to *E* the new line will follow the moves accordingly.

The new point *F* is drawn slightly darker. This means that it is not possible to move *F* freely in the move mode. *F* is a "dependent" point. The other points in a construction are "free."

Two things are worth mentioning: With the same procedure you could also have added a "half-free" point that is bound to lie on *one* line, you just had to release the mouse when you were over one line. All these operations of creating intersections, free or half-free points work similarly in the "Add a point" mode. Many other modes, the *interactive* modes, also come with this feature. Browse the reference part [p. 44] of this manual for further information.

3.1.7 Finishing the Drawing

Now it should be easy for you to finish the drawing by adding four more lines. The desired final configuration is shown in Figure 7. *First add a line from A to the intersection of b and d. Then draw two lines from A to D and from B to E. Finally, draw a line from C to the intersection of e and f.*

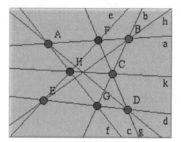

Fig. 7: Pappos' Theorem

Now you have added eight points and nine lines. If you look at the configuration you will notice that, if everything was done correctly, the lines *g*, *h* and *k* meet in a point. That this will always be the case in such a construction is Pappos' Theorem. *Switch to the move mode [p. 51] by clicking* in the tool bar, and *drag around the free points of the construction to convince yourself of the truth of this theorem.* This theorem was already known to the old Greek geometers. Later it turned out to be of fundamental importance for the theory of *Projective Geometry*.

3.1.8 Selecting and Changing the Appearance

You may not like the appearance of the drawing. The lines may be too thin and perhaps you want to emphasize the final conclusion of the theorem. *Choose the menu item "Properties/Edit Appearance".* The following window will pop up:

3.1.8 Selecting and Changing the Appearance

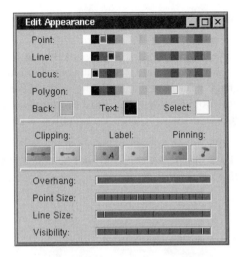

Fig. 8: The Appearance Editor

Changes in the Appearance Editor [p. 103] are immediately applied to selected elements. First *press the button* "Select all lines" [p. 50]. If you now *move the slider "Line Size" in the Appearance Editor to its second tick*, all lines become thicker. After this, *switch to "Select" mode* [p. 53]. Now, when you click over one or more elements these elements will be selected. If you hold the shift key while you click, the selection state of the element will be toggled.

*Click over line **k**, then hold the shift key and click over lines **h** and **g**.* The lines *g*, *h* and *k* should be highlighted. *Click the red box in the second row of the Appearance Editor's color palette.* The color of the three lines changes from blue to red. You can also change the size of these lines by moving the "Line Size" slider to its third tick. Finally *deselect everything [p. 50] by clicking the button* .

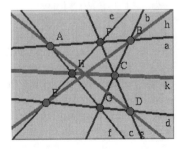

Fig. 9: Changed appearance

3.1.9 Adding a Final Point and Proving a Theorem

Before you continue open *Cinderella's* Information Window by choosing the menu item *"Views/Information Window"*. A console window pops up. *Cinderella* uses this console to report non-trivial facts about a configuration.

We will now create a final point at the intersection of the three red lines. It is hard to use the "Add a point" mode there, since *Cinderella* refuses to put a point at the intersection of three lines, because the situation is ambiguous. However, you can add a point if you *change to the "Meet" mode* [p. 76]. This mode expects you to mark two lines. A new point will be added at the intersection of these two lines. *Click two of the red lines, say **g** and **h***. The new point will be added.

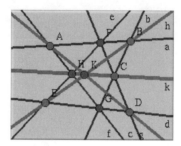

Fig. 10: The point of intersection

Notice that the newly added point **K** is always incident to line **k** due to Pappos' theorem. The console reports this remarkable fact automatically.

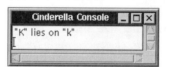

Fig. 11: A proven theorem

You may wonder how this "proving" works. *Cinderella* does *not* use symbolic methods to create a formal proof, but a technique called "Randomized Theorem Checking." First the conjecture "It seems that line **k** always passes through point **K**" is generated. Then the configuration is moved into many different positions and for each of these it is checked whether the conjecture still holds. It may sound ridiculous, but generating enough (!) random (!) examples where the theorem holds is at least as convincing as a computer-generated symbolic proof. *Cinderella* uses this method over and over to keep its own data structures clean and consistent.

3.1.10 Moving Points to Infinity

Let us explore the symmetry properties of Pappos' Theorem. For a nicer picture we want to get rid of the line labels and make the points and lines a bit smaller.

First select all lines using ![icon] *[p. 50]. Turn off the labelling by pressing the corresponding button in the Appearance Editor. Change the line size to "thin" (use the slider). Then select all points using* ![icon] *[p. 49]. Make them smaller by using the "Point size" slider. Now choose the menu option "Views/Spherical view".* What you get will not look very instructive at first sight.

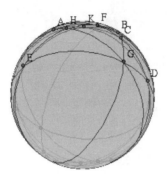

Fig. 12: A spherical view

It is a central projection of the plane to the surface of a ball. The projection point is the center of the sphere (see Figure 13). Every point is mapped to an antipodal pair of points, and each line is mapped to a great-circle (an equator) of the ball.

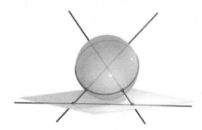

Fig. 13: Projection from the plane to the ball

In the spherical view there is a little red slider for controlling the distance of the ball to the original plane. Moving this slider corresponds to a kind of "zooming" operation on the sphere. Please *move the slider from its original position much more to the right.* Then the situation on the ball should become clearer. If you have adjusted the slider correctly, it should somehow look like Figure 14. You will clearly recognize the construction now drawn on the surface of this ball.

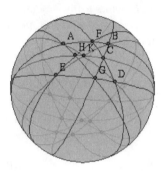

Fig. 14: Pappos' Theorem on a ball

The picture needs some explanation: For each point in the plane consider the line spanned by the point and the center of the ball. The intersection of this line with the surface of the ball gives the pair of antipodal points. For each line consider the plane spanned by the line and the center of the ball. The intersection of this plane and the surface of the ball is the great circle that represents the line in the spherical view.

On the sphere there are points that have no correspondence in the Euclidean plane. If the ball touches the Euclidean plane at the "south pole," then the equator of the ball corresponds to the "points at infinity" in the usual Euclidean plane. In **Cinderella** it is possible to make manipulations in any currently open view. Therefore you can move points also in the spherical view. The changes will be instantly reported to the Euclidean view. In particular, you can grab a point in the spherical view and move it to infinity.

We will do that now, in order to observe that our configuration has a nice three-fold symmetry. *Pick point **A** (in the spherical view) and move it to the 11 o'clock position of the boundary.* This point is now really located "at infinity." Notice that in the Euclidean view the lines passing through **A** became parallel: "The parallels meet at infinity". In a similar way, *move point **E** to the 7 o'clock position and move point **C** to the 3 o'clock position.* Your spherical picture should then look like the ball in Figure 13. In the Euclidean view you find three bundles of parallel lines. You see a kind of Euclidean specialization of Pappos' Theorem. If you like you can try to figure out the corresponding statement in terms of parallels and incidences.

3.1.10 Moving Points to Infinity

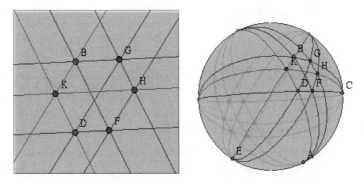

Fig. 15: A three fold symmetry

It may happen that the drawing in the Euclidean view is too big or too small. You can use the zooming tools to change this. There is a "Zoom-in" [p. 95] and a "Zoom-out" [p. 95] tool below the Euclidean view. With a press-drag-release sequence you mark the region to (from) which should be zoomed in (resp. out).

3.2 A Three Bar Linkage

In the second tutorial we want to explore the dynamics of a little mechanical linkage. In particular, you will learn how to make bars of a fixed length. You will also learn how to produce a geometric locus of a point.

3.2.1 Making a Bar

After starting **Cinderella** or after erasing your previous configuration you *switch to the "Add circle by radius" mode [p. 65] by pressing the button* 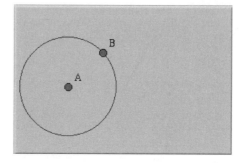 . This mode is used for circles whose radii remains constant when their centers move. *Move the mouse pointer over the view. Press the left mouse button. Hold the button while dragging the mouse. Release the mouse.* With these actions you generated a circle. Switch to the move mode to see how the circle behaves. When you pick the center you can move the position of the circle while the radius stays constant. You can also pick the circle itself and move it. Then its center stays fixed and the radius of the circle changes.

Fig. 1: A first circle

We now want to add a point that is bound to the boundary of the circle. *Select the "Add a point" mode. Move the mouse over the view. With the button pressed move the mouse pointer to the boundary of the circle. When the circle is highlighted, release the mouse.* You have now created a point that is on the boundary of the circle. You could alternatively have clicked the mouse over the circle boundary. However, in this way you have a bit less control on whether you actually hit the circle. Your configuration should now look approximately like that in Figure 1.

Switch to the "Move" mode to see how this new point behaves. If you pick it and drag the mouse it will never leave the circle. Under this restriction it tries to be as close to the mouse position as possible. When you move the center of the circle, or change the circle's radius, the point also stays on the circle. The point furthermore keeps its relative angle to the center.

Choose the interactive "Add a line" mode and draw a line from the center to the perimeter point. Open the Appearance Editor. Select the line and clip it by pressing the ▬ *button [p. 105] in the Appearance Editor window* (see Figure 2).

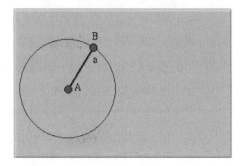

Fig. 2: A bar

This line now behaves like a bar of fixed length, which is given by the radius of the circle. The only way to change its length is to change the radius of the circle.

3.2.2 Adding Two More Bars

We now want to add two more bars to complete the three-bar-linkage that you have seen in Section 2.1.2 of the introduction. The linkage is a chain of three consecutive bars which should be pinned down at both of its ends. *Switch to the "Add a circle by radius" mode again and add a second circle right to the first one. Do not let the two circles intersect.*

The radius of the second circle represents the length of the third bar in the row (we have not created the second one, though). To finish the construction it is worth analyzing the situation. If point **C** in our construction should be linked to point **B** of the construction by two bars of given lengths, then there is not much freedom for the point that is common to these two bars. Actually, there are exactly two positions for this point. They are the two intersections of two circles: one circle around **C**, already drawn, and another circle around **B**.

24   A Quick Start

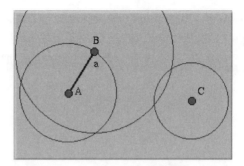

Fig. 3: Another two circles

So, as a next step *draw a circle around **B** in the "Add a circle by radius" mode. Make sure that its radius is large enough so the circle also intersects the circle around **C***. This stage of the construction is shown in Figure 3. Finally, draw a line with the "Add a line" mode from ***B*** to one of the intersections of the circle around ***B*** and the circle around ***C***.

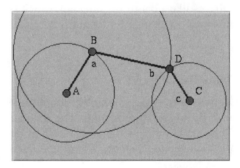

Fig. 4: All three bars

This new point is automatically labelled ***D***. Notice that this point is not a free point, since it position is determined, up to choosing the second intersection, by the positions of ***A***, ***B***, and ***C***, and the radii of our circle.

*Finish the construction by adding the third bar from **C** to **D***. Most probably the lines are already clipped, since you made this choice in the Appearance Editor. If not, select them and use the Appearance Editor to clip them. Your drawing should now look like Figure 4.

3.2.3 Moving the Construction

The lengths of the bars are determined by the radii of the circles. *Select the move mode again and play with the construction.* An interesting thing happens when you move point **B**, the point on our first circle.

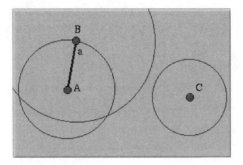

Fig. 5: The bars are too short

First of all, you notice that the whole construction has exactly one degree of freedom when moving point **B**. Moving it lets the whole construction behave as if it were a mechanical linkage. We can consider the possibility of moving **B** as a kind of "driving input force"; this is what CAD people would call it. There are positions for point **B** for which the bars are too short (see Figure 5), which means that the two circles no longer intersect. When this happens, the lines and the intersection point disappear.

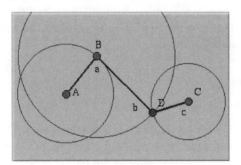

Fig. 6: After coming back

One of the most interesting things, that you might initially overlook, happens if you come back with point **B** to the old position where the circles intersect: the other point of intersection is chosen; the linkage takes the other possible position. Try this many times to get a feeling for it. *Move back and forth with point **B**, so* *that the bars disappear and appear again.* This behavior seems a bit counter intuitive at first, but it is exactly the right thing to happen. Imagine that the bars were made from real matter, steel or wood, for instance. Then they would have a certain mass. What would happen if you push the whole construction and let it move freely, driven only from inertial forces? The two bars, those that disappear in our drawing, will come to a position in which they are on a line. Then point *D* will "sweep over" and the whole construction moves to the other possible situation.

3.2.4 Starting an Animation

If you still do not believe that this is natural behavior, *choose the "Animation" mode by pressing* ⌬ *[p. 87]*. The message line below the view asks you to choose a "moving element." In our case we want to *B* to be the moving point: *Click B*. Since point *B* has only one degree of freedom it is clear how to move it, and the animation is started immediately. Otherwise you would have been asked to select also a "road" on which *B* should move. In the construction the road is clear, it is the first circle.

An animation control window pops up. It has buttons like a CD-player to start, stop and pause the animation, and it has a slider to control the animation speed. In particular, it has a button "Exit" ⌬ [p. 87] to leave the animation mode.

For a moment enjoy watching the animation. Observe that point *B* only moves in a region where the bars are still long enough. It "knows" when to change the direction. ***Cinderella*** tries to model true physical behavior. It does not add masses to the moving elements, but it uses methods from complex analysis to find out what is the most reasonable thing to do.) Please exit the animation now by pressing the "Exit" button in the Animation Controller. The Animation Controller disappears and the animation stops.

3.2.5 Drawing a Locus

We now want to know how the midpoint of the middle bar moves during the animation. We first add the midpoint. *Choose the mode* ⌬ *[p. 67]*. By a press-drag-release sequence you can add the midpoint of two points: *Move the mouse over point **D**. Press it, and with the left button pressed move over point **B**. Now release the mouse button.* The midpoint is added.

Now, *choose the "Locus" mode* ⌬ *[p. 85]*. This mode requires that you select a "mover", a "road" and a "tracer", in this order. The mover is the element that moves (the "driving force"). The road, which is literally the *road* along which the mover will move, must be incident to the mover. The tracer is the point whose trace will be calculated.

3.2.5 Drawing a Locus

*Click point **B** (the "mover").* ***Cinderella*** recognizes that this point has a unique road and selects the circle as well. So finally *point **E** has to be selected as tracer.* It takes a second and the locus is generated automatically. You can switch to the "Move" mode and see how the locus changes when youe move the free elements.

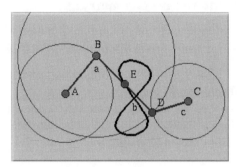

Fig. 7: Constructing a locus

If you finally *select the "Animation" mode again and click on the locus*, you will see how the locus was generated.

4 Behind the Scenes

4.1 Problems in Interactive Geometry

How should a system for doing interactive geometry behave when a user interacts with it? In a sense the requirements are similar to the requirements for other programs:

- The program should be easy to use,
- the user should not be disturbed by unnatural behavior of the program,
- the user should not be bored by being forced to make unnecessary input,
- the computed results should be correct.

Unfortunately, under these requirements interactive geometry turns out to be a difficult subject. There are two main reasons for this:

- There are problems that come from *special cases* that already occur in a static setup.
- There are problems that are of a *genuinely dynamic* nature.

4.1.1 Static Problems

Our usual "everyday geometry" is full of special cases. Two lines can intersect or they can be parallel. Two circles can intersect in one or two points or not at all. So, even for static constructions it is sometimes difficult to figure out what a correct and reasonable result for such a special case is. For instance, what is *the* angular bisector of two parallel lines? Is it undefined? Can it be any line parallel to the lines? Should it be a line that is equidistant to the two lines?

We could try to exclude all the special cases and not consider them at all. However, on the one hand this would mean excluding non-esoteric cases such as parallel lines. On the other hand, when we allow points to be moved in a construction it happens all the time that dependent elements are forced into special cases. Observe that this is still a static problem!

These kinds of static problems were studied for a long time. The great geometers of the nineteenth century were aware of them, and it is due to their effort that most of them could be solved. The key to a solution is to gradually extend Euclidean Geometry to a larger setup. First the usual plane is extended by elements at infinity, leading to *Projective Geometry*. Then the underlying algebraic structure is extended to cover *complex numbers*. This essentially removes all special cases from geometry.

It was an exciting development of mathematics how these approaches finally, around 1870, lead to a completely consistent system. This system explains the effects of euclidean geometry as well as the effects of non-euclidean, for instance hyperbolic, geometry. Today it is called "Cayley-Klein geometry" in memory of two of its main contributers.

The mathematical background and the implementation of *Cinderella* rely on this general setup. In this way *Cinderella* can deal with all special cases, and as an additional benefit is able to do non-euclidean geometry as well as Euclidean Geometry. It is an amazing fact that by using these general principles the program does not become more complicated. On the contrary: The exclusion of special cases allows a much simpler "straightforward" program structure.

4.1.2 Dynamic Problems

For systems of interactive geometry there is a second class of problems, which are in a sense more subtle than the static problems. Unfortunately they lead to even more drastic effects. Assume you have done a construction that involves points, lines and circles, and in particular the intersection of two circles (or of a circle and a line). While you move the mouse the program has to decide, for every position, where the dependent elements are. However, there is a problem. Two circles do not have only one intersection - they have two, and we get both from our calculations. How should the system decide which of them is the one you "want"? When you construct the intersection the answer to this question is easy: "Take the point that is closest to the current mouse position." But when you start to move the problem there is no immediate answer.

What would be the most desirable is a *continuous* behavior of the program in the following sense:

> "If you make a very small move with a free point, then the elements that depend on it should also move only a little bit."

At first sight it is not clear whether this requirement is satisfiable in general. Turn on your favorite system for doing interactive geometry or parametric CAD and make the following experiment: Draw a horizontal line and construct two circles of equal radius whose centers can only slide along the line. Move them to a position where they intersect and construct the upper point of intersection of the two circles. Now move one circle so that its center passes through the center of the other circle. Most probably you will see that the point of intersection suddenly jumps from the upper intersection to the lower one; it happened in all the systems we tried so far. Such a behavior counters our requirement of continuity: You make a small move, and a dependent point suddenly jumps.

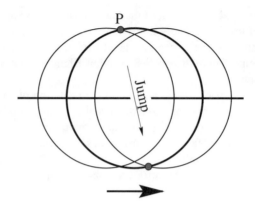

This should not happen

At first a single jumping point seems to be a kind of curiosity that is tolerable. But what happens if large parts of a construction depend on this jumping point? Then these parts of the construction will jump too, without prior warning. Most systems for interactive geometry use heuristics based on orientation decisions that help to get rid of some of these jump situations. But still in every system many cases remain unresolved. Actually, there is a proof that no heuristic based on orientations only will be able to resolve all of these dynamic problems [Kor, KRG]. In an article on dynamic geometry [Lab] Jean-Marie Laborde, the main designer of Cabri Géomètre, states this dilemma in the following way:

> "I think we need a real mathematical treatment of all consequences of stretching geometry in some way to a wider (dynamic) system. This system cannot be the projective one if we want to maximize the way the environment takes into account the special characteristics of non-static objects which are at the core of dynamic geometry."

Cinderella is the first program that is based on a theory which is capable of preventing dependent elements from jumping. This theory is also based on the use of *complex numbers*, which were used to solve the problems of static geometry.

The use of this theory has many benefits. For instance, it is the basis of the generation of correct loci. Consider the "three-bar-linkage" example of the second tutorial. The generation of the locus is based on the correct calculation of the points of intersection of two circles while a free point moves. In other systems for interactive geometry you will probably only get half of the eight-curve. The methods for automatic theorem proving, which are used internally throughout **Cinderella**, are based on this theory, too.

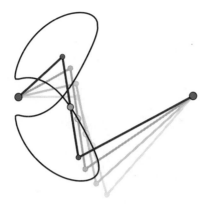

The "three-bar locus" revisited

The following pages should give you an impression of the different mathematical methods and theories that form the basis of the implementation of *Cinderella*.

4.2 Projective Geometry

The first and perhaps most important step for a consistent geometric setup is to extend the usual Euclidean plane to contain elements at infinity. You have surely heard the phrase "parallel lines meet at infinity", and you might believe it when looking from a bridge along a very long and straight railroad track. This phrase is the key to "Projective Geometry." The extension of geometry by infinite elements removes a lot of special cases from usual Euclidean Geometry.

Projective Geometry has a very long tradition. Its historical origin traces back to the study of perspective undertaken by famous painters such as Albrecht Dürer and Leonardo da Vinci. Its mathematical origin is the work of *Gaspard Monge*, a French geometer, who developed, around 1795, a method called *descriptive geometry* for representing spatial configurations by planar perspective drawings. Monge observed that non-trivial facts about planar geometric configurations could be derived by considering these configurations as projections of configurations in space. The study of parallels in these projections was most elegantly done by extending the plane by elements at infinity.

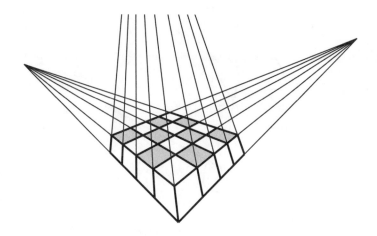

In perspective drawings parallels actually meet

The *projective plane* consists of the points of the usual, Euclidean, plane together with one additional "infinite" point for every possible direction. The lines of the projective plane are the Euclidean lines together with one special "line at infinity." All infinite points lie on the line at infinity. The following, nicely symmetric, relations between points and lines hold:

- Any two distinct points have a unique connecting line (their "Join")
- Any two distinct lines have a unique point of intersection (their "Meet")

The first person who formalized these rules, this was around 1822, was *Victor Poncelet*, a student of Monge, who today can be called the "father of Projective Geometry." In Projective Geometry there is no need to consider parallels as something special. They still have a point of intersection, it just happens to lie at infinity. For a readable introduction to Projective Geometry we refer to the books of H. S. M. Coxeter on that topic [Cox1, Cox2].

4.3 Homogeneous Coordinates

On a computer we unfortunately do not have geometric objects as primitive data types. A point or a line has to be represented by numbers: the coordinates. Usually a point in the plane is described by its *(x,y)*-coordinates. A line may be given by the three parameters *(a,b,c)* of its defining equation $ax + by + c = 0$. However, when we want to do Projective Geometry this turns out to be impractical. Each point coordinate *(x,y)* represents a finite point and there is no representation for the points at infinity. The correct solution to this problem became gradually clear in the first half of the nineteenth century. It started with Möbius' *barycentric coordinates*, via the refined setup of *homogeneous coordinates* given by Plücker, and finally lead to Grassmann's setup of *multilinear algebra*.

The way out of the dilemma is as follows. For every point we use three instead of two coordinates, introducing a third dimension. Consider the following scenario: The plane is embedded parallel to the *x,y*-plane of three-space at a height of $z = 1$, so it does not pass through the spatial origin.

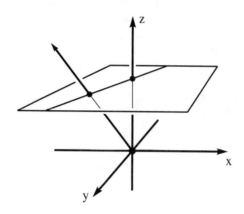

Embedding the plane in space

Every point *(x,y)* is represented by its three-dimensional coordinates *(x,y,1)*. These coordinates are the *homogeneous coordinates* of the point. What happens with the rest of the points in the three dimensional space? Almost all of them will be interpreted as points in the original plane: we identify all three dimensional points that differ by a non-zero multiple. For instance, *(4,6,2)* and *(2,3,1)* describe the same point. In general, a point with spatial coordinates *(x,y,z)* is identified with the point *(x/z,y/z,1)* of the original plane. This process is called de-homogenization. In a way every point of the original plane corresponds to the line spanned by this point and the origin in three-space.

However, there are points in three-space that do not correspond to points in the original plane. The points of the form *(x,y,0)* cannot be de-homogenized in the above way, since then we would have to divide by zero. These points correspond precisely to the "points at infinity" of Projective Geometry. To see this, we study the behavior of a point that gradually moves to infinity in the original plane.

Assume that the moving point has coordinates *(r,r)*. When *r* becomes larger and larger this point gradually approaches a point at infinity in the 45° direction. Looking at its homogeneous coordinates we see that they have the form *(r,r,1)~(1,1,1/r)*. When *r* increases, the contribution of the first two coordinates dominate the last coordinate. In the limit case where *r* equals "infinity" the homogeneous coordinates are given by *(1,1,0)*, an infinite point. You can also try to imagine the line through this point and the three dimensional origin. When the point approaches infinity this line becomes more and more horizontal, until, in the limit case, it is entirely contained in the *x,y*-plane.

A similar representation can be given for lines. For the line *ax + by + c = 0* we take the parameters *(a,b,c)* as the homogeneous coordinates of the line. As in the case of points we identify non-zero multiples of such coordinates, since they do not alter the solution space of the corresponding equation. There is one set of parameters *(0,0,1)* that does not correspond to a finite line. This is the line at infinity. The vector *(a,b,c)* of a line is orthogonal to the plane spanned by the corresponding line and the origin of three-space. In particular the vector *(0,0,1)* is orthogonal to the *(x,y)*-plane, the "line at infinity."

In fact, the algebraic notion of homogeneous coordinates provides a complete symmetry between points and lines. Each point or line is represented by three homogeneous coordinates. A point *(x,y,z)* lies on a line *(a,b,c)* if and only if the scalar product *ax + by + cz* is zero - this is nothing but rewriting the equation of the line. Geometrically this means that the two corresponding vectors are orthogonal in three-space.

4.4 Complex Numbers

Not only geometry was extended throughout the centuries. A similar process happened to *numbers*, too. Probably the first numerical concept considered by mankind were the integers: *1, 2, 3, ...*. Starting from there it was reasonable to gradually extend the system of integers to more powerful concepts. The *negative numbers*, the *rational numbers* and the *real numbers* had to be invented to get a useful and self-contained system. The observation that there must be numbers that cannot be represented as fractions of two integers is of geometric nature and dates back to approximately 600 BC. It was seen by the Pythagoreans that there is no rational number measuring the length of a diagonal of a quadrangle with sides of length one. Applying Pythagoras' Theorem this task is equivalent to finding a number *x* with $x^2 = 2$. This discovery lead to a deep crisis in the foundations of ancient geometry.

However, the story of extending the number system does not stop at that point. One of the extensions, with perhaps the most drastic consequences, was the introduction of *complex numbers*. It was *Geronimo Cardano* in his *Ars Magna* that appeared in 1545 who was first to explicitly propose such an extension of the reals. He was lead to his conclusions by the study of solutions of polynomials of degree three. Based on the work of other contemporary mathematicians he discovered that a complete systematic representation of these solutions can only be given with the help of hitherto unknown values.

A complex number is a number of the form *a + i·b* where *i* satisfies the equation $i^2 = -1$, and *a* and *b* are real numbers. Clearly the number *i* cannot be any real number since the square of a real number can never be negative. The system of complex numbers is, just like the real numbers, closed under addition and multiplication. In other words the sum and the product of two complex numbers can again be written in the form $a + i \cdot b$ for suitable parameters a and b. However, unlike the real numbers the system of complex numbers is also closed under the operation of finding solutions of polynomials. For instance, consider the polynomial:

$$x^2 - 6x + 13 = 0.$$

As you can easily check it has no real solutions, but the complex numbers $3 + 2i$ and $3 - 2i$ do solve this equation. In fact, the following beautiful result is true: *Every polynomial with arbitrary real or complex coefficients has all its solutions again in the field of complex numbers.*

In a sense the discovery of complex numbers is the starting point for most modern mathematics. Many mathematical theories find their broadest, most elegant and most economic setting when they are formulated over the complex numbers. This happens also to geometry. Consider the situation of two circles. Depending on their position they can have *two*, *one*, or *no* points of intersection.

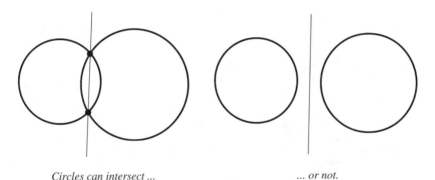

Circles can intersect ...   *... or not.*

Finding the coordinates for the points of intersection is nothing else but solving a quadratic equation. Over the real numbers this equation can have no solution. In this case the circles do not intersect. Over the *complex numbers* a solution always exists. So we can say that in the case of visually non-intersecting circles the intersections still exist: *they have complex coordinates and we cannot see them in the real plane.*

Cinderella's mathematical kernel is implemented entirely over the complex numbers. So when intersections visually vanish **Cinderella** does not have to deal with special cases, and it can still continue calculating - the solutions just have complex coordinates.

What happens if two complex points are connected by a line? In general, this line will also have complex coordinates. However, if the points are so-called *complex conjugates*, which means that they differ only by the sign of their complex part, then their join is again real. It is the case that the two intersections of two circles are always complex conjugates. This is the reason why the line joining them is a real line, no matter where the circles are. **Cinderella** will correctly calculate and draw this line, independent of the position of the circles. It may take a while to get used to the fact that intermediate results can disappear while some constructions depending on them remain visible. However, this is exactly what you should expect. Consider the case where the circles have the same radius. The line is then the *perpendicular bisector* of the segment joining the two midpoints. If you include the complex situations, you just have to consider less special cases.

Another example for a theorem where intermediate steps disappear is the following statement about three circles. Construct the line joining the two points of intersection of each pair. The three chords that you get this way meet in a point - no matter whether the circles intersect or not.

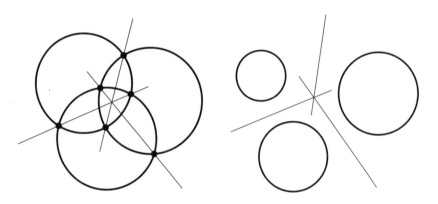

Intermediate point can be complex ... ... *but theorems stay true!*

So in **Cinderella** each point and each line is represented by *complex homogeneous coordinates*. This means that altogether any point or line has a six-dimensional (!) internal representation in the mathematical kernel. This may sound crazy, but it is the most natural thing to do.

4.5 Measurements and Complex Numbers

If we were satisfied with projective incidence theory then the system presented so far would be fairly complete. However, we want to be able to measure distances and angles, too. Measurements are in a sense the most fundamental geometrical operations. Unfortunately, Projective Geometry is not capable of measuring at first sight, since under perspective transformations distances can change. Actually, for a long time mathematicians considered Projective Geometry a "nice toy" for doing incidence geometry, but not appropriate for the real stuff: *measuring*.

History proved them wrong. With the right setup Projective Geometry is *the* universal system for doing measurements. This system unifies and explains different kinds of measurements, for example, it explains the relationship between Euclidean and hyperbolic geometry. However, it took a long time to finally find the algebraic setting in which Projective Geometry develops its full power. The key objects are called "Cayley-Klein Geometries" in modern terms. It is an elegant and consistent mathematical approach to measurements that combines Projective Geometry and complex numbers.

4.5.1 Euclidean and Non-euclidean Geometry

One part of this development started with the discovery of non-euclidean geometries. Our everyday geometry is, with relatively great accuracy, described by Euclid's five postulates. He used these postulates to axiomatize geometry - this happened almost 2000 years ago. The last postulate, the so-called "Parallel Axiom", plays a special role in the development of geometry. A way to formulate it is: *"Whenever there is a line l in the plane and a point P not on l, then there is **exactly one** line through P that does not meet l."*

Euclid was very cautious about using the Parallel Postulate. Large parts of Euclid's elaborations, for instance, the complete theory of congruence of triangles, were done without the explicit need for this axiom. Today we are relatively sure that Euclid himself believed that this axiom was a consequence of the other four axioms. But he could not prove this. After Euclid many other mathematicians tried to do so, some of them even presented proofs. But all these proofs were incorrect.

In the 16th to 18th century mathematicians also found many equivalent formulations for the Parallel Postulate. One of the most prominent formulations is *"The inner angles in a triangle sum up to $180°$."* If this statement could be derived from Euclid's first four axioms then this would prove the dependence of the Parallel Postulate.

Proving that an axiom is dependent can be done by assuming its contrary and drawing conclusions until a contradiction is shown. Many people tried this, among them *C.F. Gauss*, *J. Bolyai* and *N. Lobachevski*. They drew conclusion after conclusion, but, to their surprise, instead of arriving at a contradiction they found themselves developing a beautiful geometric system: *hyperbolic geometry*. There

Euclid's Parallel Postulate is modified in the following way: *"Whenever there is a line **l** in the plane and a point **P** not on **l**, then there will be more than one line through **P** that does not meet **l**."* A consequence of this assumption is that the angle sum in a triangle is always less than 180°. Between 1815 and 1824, independently from each other, these three people - who are today considered as the discoverers of hyperbolic geometry - came to the point at which they declared their system as free from contradictions, just because they could not find any. The system they developed was full of inner beauty, and it is a surprising fact that they could prove that under the assumption of the perturbed Parallel Postulate they end up with an up to trivial isomorphisms unique theory.

It is worth mentioning that most probably Gauss was the first who arrived at these conclusions, around 1816. However, he did not dare to publish his results, since he was afraid of conflicts with the leading schools of Kantian philosophy at that time. They considered a straight line as the first thing whose nature is "a priori" clear.

If you are interested in the history of mathematics we want to point you to the books of Bell [Bel1, Bel2] and Struik [Str]. As an introduction to hyperbolic geometry we recommend the book of M. J. Greenberg [Gre].

4.5.2 Cayley-Klein Geometries

For a long time it was not clear whether the system of hyperbolic geometry was indeed free of contradiction. What was missing was a model of this structure, a mathematical object that satisfied Euclid's first four axioms and the perturbed Parallel Postulate. With minor flaws Beltrami was the first who could construct such a model in 1868. However, the full beauty of a general theory was first seen when Felix Klein, student of Plücker, presented his version of what we call "Cayley-Klein geometries", see for instance [Kl2]. What he gave was essentially a reduction of hyperbolic geometry to constructions of Euclidean Geometry that implies: "If Euclidean geometry is free of contradictions then so is hyperbolic geometry." This finally solved all problems around Euclid's fifth postulate.

The idea behind Cayley-Klein geometries is to use the Projective plane, and to distinguish a special conic - the "fundamental object." A special kind of global measurement is defined that depends only on the fundamental object. Depending on the type of the fundamental object you have chosen you get different types of geometries: *Euclidean geometry, hyperbolic geometry, elliptic geometry, relativistic geometry*, and three other geometries of minor importance.

We will not go into the details of Cayley-Klein geometries, but we will present the major definitions and demonstrate some basic effects. We first need the concept of a *cross ratio*: For four points A, B, C and D on a line, the cross ratio is defined as the number

$$(A,B \mid C,D) := ((A\text{-}C)(B\text{-}D)) / ((A\text{-}D)(B\text{-}C))$$

where *(A-C)* denotes the usual "Euclidean distance" of the points *A* and *C*. The cross ratio can also be defined without referring to the notion of Euclidean distance, which is important for a systematic treatment of geometry that is free from circular conclusions.

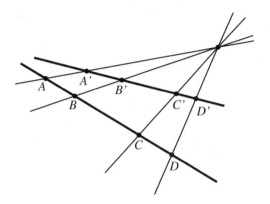

The cross ratio

The cross ratio is of remarkable value in Projective Geometry, since it is invariant under perspective transformations. So, if you have four points *A*, *B*, *C* and *D* on a line and centrally project them to four points *A'*, *B'*, *C'* and *D'* on another line, then the cross ratios of the two quadruples of points are identical.

Similarly, the cross ratio of four lines through a point *P* can be defined to be the cross ratio of four points that are the intersections of each line with another, distinct line not going throught the point *P*.

Now the definition of a Cayley-Klein geometry is easy. Pick a quadratic form

$$ax^2 + by^2 + cz^2 + dxy + exz + fyz = 0.$$

The zero set of this equation describes a, possibly complex, conic in the projective plane. This is the "fundamental object" of the geometry. Now measurements of angles and distances are defined as follows: For the distance between two points *A* and *B* take the line joining them. The intersection of this line with the fundamental object are two points *X* and *Y*. Calculate the cross ratio *(A,B | X,Y)*. Take the logarithm of that number and call the result "distance."

Angles are calculated in an analogous way. For the angle between two lines *L* and *M* first take their meet, that is their point of intersection. The tangents through the meet that touch the fundamental object are two lines *P* and *Q*. Calculate the cross ratio *(L,M | P,Q)*. Take the logarithm of that number and call the result "angle." Usually, these two functions are multiplied with some cosmetic constants *r* and *s* in order to match the traditional definitions of measurements.

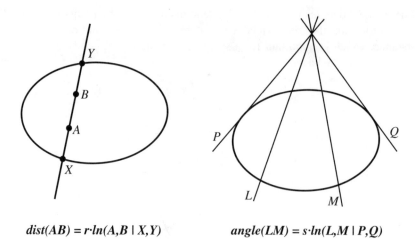

$$dist(AB) = r \cdot ln(A,B \mid X,Y) \qquad angle(LM) = s \cdot ln(L,M \mid P,Q)$$

It may sound like magic, but this is all you have to know. Depending on the type of fundamental object that you have chosen you get different kinds of geometry. Up to isomorphism there are exactly seven different types of geometries you obtain that way. The three most important choices for the fundamental conic are:

1. The circle given by $x^2 + y^2 - z^2 = 0$.
   The resulting measurement corresponds to hyperbolic geometry.
2. The degenerate conic described by $x^2 + y^2 = 0$.
   The resulting measurement corresponds to usual Euclidean Geometry.
3. The equation $x^2 + y^2 + z^2 = 0$ with no real solutions.
   The resulting measurement corresponds to elliptic geometry.

Two things are worth mentioning:

- Distances in Euclidean Geometry need a little twist. The formulas of Cayley-Klein geometry will always produce a zero distance. This is due to the fact that in Euclidean Geometry there is no "absolute" notion of distance. Each length has to be compared to a unit length. The right formulas are obtained immediately in the limit situation.
- The intersections and tangents in the above constructions do not necessarily have real coordinates. For instance, in the case of elliptic geometry the fundamental object has no real points at all. In this case the intersections of the fundamental object with a line are always complex.

The metric part of **Cinderella** is based on Cayley-Klein geometries. All calculations of lengths, angles, orthogonality, circles, etc. refer to a fundamental object.

We finally want show at least one effect that is caused by this general theory. We do this in order to give you a feeling for what complex numbers, cross ratios and Projective Geometry have to do with measuring.

We consider the case of Euclidean Geometry. There the fundamental object has the equation $x^2 + y^2 = 0$. Using complex numbers this quadratic form can be factored into two linear forms: $x^2 + y^2 = (1 \cdot x + i \cdot y + 0 \cdot z) \cdot (1 \cdot x - i \cdot y + 0 \cdot z)$. The points $I := (1,i,0)$ and $J := (1,-i,0)$ that occur in this formula play a special role in Euclidean Geometry. They are not affected by any euclidean transformation. In a very well defined way we can say that: *"Euclidean Geometry is Projective Geometry together with I and J."*

The points I and J are sometimes called the *imaginary circle points*, since they have a very special relation to circles: each Euclidean circle passes through I and J. To see this consider a general circle equation in homogeneous coordinates

$$x^2 + y^2 + cz^2 + exz + fyz = 0.$$

Now plug in the coordinates of I and J. Using the rules for calculating with complex numbers we observe that the circle equation is obviously satisfied. Thus we can say that a circle is a special conic that passes through I and J. With the notion of a circle it is easy to define what it means to have equal distances or equal angles. The remaining concepts of Euclidean Geometry can be derived in a straightforward manner.

4.6 The Principle of Continuity

It was mentioned in the preface of this manual that **Cinderella** uses some basic new methods to avoid inconsistent behavior. The geometric system we presented in the previous sections is a closed framework for doing geometry, including measurements. However, so far there is a element missing that is crucial for **Cinderella**: *dynamics*. Most other systems for interactive geometry, or parametric CAD, suffer from inconsistencies that come from an unsatisfactory treatment of the special effects of dynamics. For instance, consider the "theorem" that *the angular bisectors of the sides of a triangle meet in a point*. Every pair of lines has two angular bisectors which are perpendicular to each other. Depending on the choice of the angular bisectors the above statement can be true or false. Now imagine you have constructed an instance of this "theorem" (i.e. you have chosen the correct bisectors). You start to drag the vertices around and suddenly, without reason, one angular bisector flips to the other position and the "theorem" becomes false. Such a scenario can happen in a any system that does not take extra efforts to resolve the special problems from the dynamic aspects of geometry.

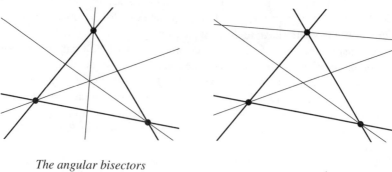

The angular bisectors of a triangle can intersect ...   *... or not*

Let us consider another small construction. Take two circles and one of their intersections. While you drag elements around **Cinderella** has to decide for every mouse move which intersection "you mean."

It is a good first attempt to "trace" this point of intersection using the rule: *"always take the intersection that is closest to the previous position"*, since this precisely reflects the definition of continuity. But how should we deal with vanishing intersections? Again it pays off having implemented everything in complex space. In **Cinderella** intersections never vanish, they can become complex, though. So we have to trace the intersections in complex space and use the above rule.

However, this is not enough. When you separate two circles that were previously intersecting there is always a position in which the two points of intersection coincide. How can the points be distinguished in this situation? This time it is analytic function theory that rescues us. *If "detours" through complex space are allowed, then there is always a path that avoids all degenerate situations.* Again the whole approach is only possible since everything is embedded in a complex space.

Here is an approximate description of what happens while you drag the mouse from one position to another in the "Move" mode of **Cinderella**. While you move a point from position *A* to position *B*:

- **Cinderella** generates a path from *A* to *B* through complex space which avoids all degeneracies,
- the dependent elements are traced through complex space.
- the number of intermediate steps on the path is adjusted according to the required accuracy

For the tracing *Cinderella* uses an adaptive step width algorithm. You can imagine that while you drag the construction elements the mouse pointer leaves the "real screen" and walks through complex space.

Why all this effort? With these methods we can be sure that elements do not "jump around" without any reason. So when you start with a correct drawing of the angular bisectors theorem you will never be able to move it to a false position. The theory forms also the basis for the reliable randomized theorem checking and for the loci and animation functions of *Cinderella*.

5 Reference

5.1 Overview

During a typical working session with *Cinderella* one or more main windows can be opened. In these windows most of the actions are performed. A typical main window looks like this:

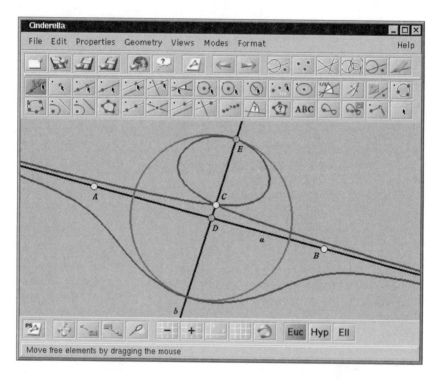

Each main window of *Cinderella* consists of six major parts. From the top to the bottom there are:

1. *The menu bar*
   From there you can access almost all actions, including file export, geometric operations and selection tools.
2. *A toolbar with general actions*
   There you have access to actions like Save, Load, Print, Export to HTML, Undo/Redo operations and various selection tools.

3. *A toolbar with geometric modes*
   This toolbar contains all major geometric operations ranging from simple intersection operations to complicated operations like the generation of loci.
4. *The drawing surface (view)*
   This is the place where all the actions take place. Here you will construct and make your geometric explorations.
5. *A toolbar with view-specific actions*
   This toolbar contains operations like Zooming, Translating, Scaling, PostScript export, etc.
6. *The message line*
   In this line **Cinderella** tells you what it is currently doing or what it expects as input from you.

5.1.1 The Menu

Almost all functionality of **Cinderella** is accessible through the menu. However, it is much more convenient to use the different toolbars to access the different operations. You can (de-)collapse toolbars by a double-click on their background, if you need more space for your construction.

The menu bar has six default entries:

- *File*
  The standard file operations.
- *Edit*
  Undo, Redo and selection tools.
- *Properties*
  Changing the appearance of the geometric objects.
- *Geometry*
  Selecting the type of geometry (Euclidean, hyperbolic or elliptic).
- *Views*
  Opening additional views of the construction.
- *Modes*
  Access to the geometric operations.
- *Format*
  Selecting the format in which coordinates of the elements are presented..

5.1.2 The General Action Toolbar

The general toolbar collects *actions* for several purposes. There you have tools for the standard file operations, export facilities, undo and redo, and various selection tools. The operations provided in detail are the following:

46   Reference

File operations: [p. 48]	new	load	save	save as	
Export tools: [p. 48]	print	export HTML	export exercise		
Undo/Delete: [p. 49]	undo	redo	delete		
Selection tools: [p. 49]	select all	select points	select lines	select conics	select none

All these actions have one thing in common: they do not rely on sophisticated mouse interaction in the view. Simply click the action and something happens (a file is saved, elements are selected, an undo is performed, etc.).

5.1.3 The Geometric Tools

Unlike the actions mentioned above the *modes* expect you to perform operations in the drawing surface. So, when you select a mode usually nothing happens immediately, the program expects additional mouse input from you. The message line tells you to exactly what **Cinderella** expects. **Cinderella** is always in a particular mode. You can easily identify the mode that is currently active, since its button is slightly darkened. A fully equipped toolbar for geometric operations looks like this:

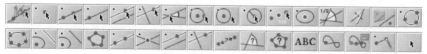

The variety of these operations enables the full geometric power of **Cinderella**. The geometric tools (modes) can be roughly grouped into six categories:

- The move mode [p. 51]
  This mode is the key feature of *Cinderella*. It allows you to drag around the base elements of a construction. While you move such an element the whole construction changes consistently.
- The select mode [p. 53]
  This mode allows you to select elements with the mouse. Selected elements may either be used as input for *definition modes* or their appearance may be changed using the Appearance Editor.
- Interactive modes [p. 54]
  These modes are very powerful construction aids. They take full advantage of mouse interactivity. A click-drag-release sequence often constructs more than one element. Moreover, the actual definitions of the new elements are adapted with respect to the actual mouse position.
- Definition modes [p. 68]
  Unlike in the interactive modes, the definition modes do not require sophisticated mouse interactivity. The construction scheme provided by definition modes is fairly simple: one selects elements by mouse clicking. As soon as enough elements for the desired definition are selected, new elements are added, and the game starts again. Usually a message line supplies you with information about the kind of elements *Cinderella* expects as input.
- Measurements [p. 79]
  These modes allow elementary geometric measurements to be determined. Similarly to the definition modes, only the defining elements have to be selected.
- Special features [p. 83]
  These modes provide special effects like loci, animations, text or line segments.

5.1.4 The Geometries

In *Cinderella* you have "native support" for non-euclidean geometries. That means that if you want to do, for instance, hyperbolic geometry, you do not have to mimic this with Euclidean constructions, you switch to "hyperbolic" mode instead. You can choose different geometries by using the *"Geometries"* [p. 91] menu entry, or by using the corresponding buttons in the view specific toolbar.

5.1.5 The Views

Cinderella offers a variety of different views in which you can observe a configuration. Each view corresponds to a specific way of representing geometry. Besides standard Euclidean views [p. 94] there are also views for hyperbolic [p. 98] or spherical [p. 96] geometry. A textual description [p. 100] is also available. You can access the different views via the *"Views"* menu.

5.2 General Tools

5.2.1 File Operations

The File Operations are rather standard and similar to corresponding operations in other programs. Constructions are stored in a special file format. The construction files have the extension ".cdy".

5.2.1.1 New

This action clears the entire construction. Everything is reset to the original state.

5.2.1.2 Load

Loads a construction.

5.2.1.3 Save

Prompts for a filename if needed and saves the construction.

5.2.1.4 Save As

Prompts for a new filename and save the construction.

5.2.2 Export Tools

5.2.2.1 Print All Views

Prints all currently open views. This operation makes use of the Java printing routines. The output is inferior to the output produced using the "Print PS" command of a view, due to the limits of the Java printing interface. If you have the facilities to print and preview PostScript files (we recommend installing Ghostscript and GSView resp. gv) it is much better to use the PostScript printout routines.

5.2.2.2 Export to HTML

This operation automatically generates an interactive HTML page.

5.2.2.3　Create an Exercise

With this operation you can generate student exercises, which can be used inside a standard web browser. Read more about it in the chapter on creating interactive web pages and exercises [p. 110].

5.2.3　Undo/Redo

5.2.3.1　Undo

This operation undoes the last performed action. The following actions are undoable:

- Construction steps
- Movements
- Appearance changes
- Zooming, translating, and rotation of views
- Deletion of elements

  You can undo as many consecutive operations as you want.

5.2.3.2　Redo

Redoes the last undo operation. You can redo as many consecutive undo operations as you want.

5.2.3.3　Delete

Deletes all selected elements and all elements that depend on them. If you delete elements by accident, just use "Undo" to restore them.

5.2.4　Selection Tools

5.2.4.1　Select All

Selects all geometric elements.

5.2.4.2 Select Points

Selects all points.

5.2.4.3 Select Lines

Selects all lines.

5.2.4.4 Select Conics

Selects all circles and conics.

5.2.4.5 Deselect

Clears the current selection.

5.2.5 Moving an Element

This mode is probably the most important feature of ***Cinderella***. It allows objects to be moved around. Usually this mode is used to move the free elements of a construction. For this place the mouse over a movable element, pick the element by pressing the left mouse button, and drag it (with the button still pressed) to the new desired position. After the button is released, another element can be moved around.

Cinderella distinguishes between two types of elements: *movable elements* and *fixed elements*. As the name indicates, movable elements are those that you can move around in this modus. The fixed elements are those whose position is already entirely determined by the rest of the construction. If, for instance, a point is defined to be the intersection of two already defined lines, then it is no longer movable and becomes a fixed element. You can visually distinguish movable points from fixed points by their appearance: the movable points are drawn brighter.

There are six types of movable elements:

- *Free Points:* Points that are not dependent on other elements of the construction. These points can be moved around freely.
- *Points on a line:* Points may be defined to be always incident to a certain line. During a move these points slide along the line.
- *Points on a circle:* Points may be defined to be always incident to a certain circle. During a move these points slide along the circle.
- *Line with a slope:* A line may be defined to be always incident to an already constructed point, with no further restrictions. Such a line can be picked directly for a move. During the move the line rotates around the point.
- *Circles by center and radius:* A circle that is defined by its center and its radius may be scaled during the move mode. You pick the boundary of the circle with the mouse and drag the mouse. The radius changes according to the actual mouse position. You can move the whole circle by dragging its center point.
- *Texts and Measurements:* Texts and measurements may be also dragged around in the move mode. You pick the text and move it to the desired position. Texts and measurements may be placed everywhere in the view. The boundary of the view has snap points that support the creation of nice layouts. Texts and measurements may be also "docked" to points of the construction.

There is one more reason to use the "move-mode": you can use it to *change the position of labels*. For this press the *control-key* of the keyboard, pick the label with the mouse and move it to the new position. However, labels cannot be moved to any position. They must be somehow close to their corresponding element. Movements that drag the label too far away are blocked automatically.

Synopsis

Move elements by dragging.

See also

- Add a point [p. 55]
- Line through point [p. 59]
- Circle by radius [p. 65]
- Add text [p. 83]

5.2.6 Select

With this mode you can select elements with the mouse. You can easily recognize selected elements, since they are highlighted in all views. There are three reasons why you might want to select elements:

- The most common use of the select mode is individualizing the appearance of geometric elements. All changes that you make in the Appearance Editor [p. 103] are immediately applied to all selected elements. If for instance several lines are selected and you move the size slider in the the Appearance Editor, then the thickness of the selected lines changes.
- Selected elements may be deleted by using the "Delete" action.
- Elements may be selected to define "Input", "Output" and "Hints" for an exercise. For details on this consult the section about creating interactive web pages and exercises [p. 110].

You can select elements in the view either by clicking over them or by dragging the mouse with left button pressed. Depending on what you do, four different behaviors are possible:

- When you click with the mouse somewhere on a view, then exactly those elements hit by the mouse pointer will become selected. All other elements will be deselected.
- When you keep the shift-key down while clicking the mouse over an element then the selection state of the element will toggle. The selection state of the elements not hit by the mouse will stay the same. You can use "shift-clicking" to select many objects.
- When you drag the mouse while holding the left button down, then those elements that you touch while you move will become selected.
- When you drag the mouse while holding the shift-key and the left button down, then those elements that you touch while you move will be deselected.

We strongly recommend playing with the different possibilities of the select mode.

Synopsis

Select individual elements with the mouse.

5.2.7 Interactive Modes

These modes are very powerful construction aids. They take full advantage of the mouse. A single click-drag-release sequence constructs several elements. Moreover, the definitions of the new elements are adapted according to the mouse position. For instance, in the "Add a line" mode, the line, together with its start and endpoints, are generated. The definition of the start and endpoint depends on the mouse position when the mouse button is pressed resp. released. So, when the mouse pointer is over an intersection of two lines, then the new point will be defined as the point of intersection of these lines.

If elements are already present in the construction, they will not be added a second time. This refers not only to the definition of the elements, but it is guaranteed by *Cinderella's* automatic theorem checking facilities.

Interactive modes can easily be recognized in the tool bar since their icon contains a little mouse pointer symbol.

5.2.7.1 Add a Point

A single new point will be added using this mode usually. The mode is designed to be multi-purpose and easy to use. Pressing the left button of the mouse generates a new point. While you drag the mouse, with the left button still pressed, you can change the position and definition of the newly added point. When you finally release the mouse button the point is fixed. Its definition depends on the position where the mouse was released. Sometimes it can be more convenient to use the more powerful interactive modes (*"add a line"* [p. 57], *"add a parallel"* [p. 60], *"add a circle"* [p. 64], etc.), which generate points together with other geometric elements.

As already mentioned, constructing a point can be described as a three step procedure:

- *Pressing the left mouse button* generates the point.
- *Dragging the mouse* moves the new point close to the mouse position. You will recognize that the point snaps to already existing elements. Whenever this happens the definition of the point is adapted to the current situation and the defining elements are highlighted. In particular, the point snaps to intersections of already existing elements, and it snaps to already existing points.
- *Releasing the mouse button*
  fixes the definition of the point. If at this moment the mouse is ...
    - ... over no element at all, then a free point is added with no additional constraints. In the "move mode" this point can be freely moved.
    - ... over an already existing line, then the new point will always be incident to the line. In the "move mode" this point can only slide along the line.
    - ... over an already existing circle, then a point that is always incident to this circle is added. In the "move mode" this point can only slide along the circle.
    - ... over the intersection of two elements (line, circle, or conic) then the intersection of these elements is added. Also such a point will appear slightly darker than the free points. It is no longer freely movable in the "move mode".
    - ... over an already existing point, then no element will be added.
      The elements that lead to the definition will always be highlighted.

The figures below show the three main situations: a "free point," a "point on a line," and an "intersection point." Notice that while dragging the mouse the coordinates that correspond to the point's position are displayed.

Free point: *Point on a line:* *Intersection point:*

Clearly a point can also be added with a simple mouse click at the position where the point should be placed. The above mentioned three step procedure collapses then to a single mouse click. However, it may be a matter of taste, it is often more convenient to take the full advantage of the three steps mentioned above:

Press the mouse button down somewhere, drag the point to the right position and release the mouse.

Synopsis

Create a new point with a press-drag-release sequence. The definition is automatically adapted.

Caution

There are two situations where it is not appropriate to add a point with this mode.

- If you want to add an intersection of two lines that cross outside the visible area of the view, then you should use the *meet* mode [p. 76].
- If you want to add a point at a position where more than two elements cross, then you should either
  - perturb the situation slightly by using the *move* mode [p. 51], if possible,
  - zoom in to get a higher resolution [p. 95]
  - or for line-line intersections use the *meet* mode [p. 76] to define exactly what intersection is meant.

See also

- Move an element [p. 51]
- Add a line [p. 57]
- Add a circle [p. 64]
- Meet [p. 76]

5.2.7.2 Add a Line

This mode allows a line through two points to be added. The mode is powerful enough to generate the line together with the two points with just one mouse action. When the mouse is pressed the first point is added. Dragging the mouse generates the second point and the line. When the mouse is released the position of the second point is frozen, and the construction is finished. The logic behind this mode is very similar to the logic behind other interactive modes (*"Add a parallel"* [p. 60], *"Add a perpendicular"* [p. 62], *"add a circle"* [p. 64], etc.).

Constructing a line in this mode is a three step procedure:

- *Pressing the left mouse button* generates the first point. The definition of this point depends on the position of the mouse at the moment when the button is pressed:
    - If the mouse pointer is over an already existing point, then this point is taken.
    - If the mouse pointer is over the intersection of two elements (line, circle, or conic), then this intersection is automatically constructed and taken as the first point.
    - If the mouse pointer is over just one element (line or circle), then a point is constructed that is constrained to this element. This point is taken as the first point.
    - Otherwise a free point is added.
- *Dragging the mouse* generates the line and the second point. The definition of the second point is chosen depending on the mouse position. As in the *"Add a point"* mode the second point snaps to already existing elements. The choice of definitions is completely analogous to the choice for the first point. The elements that define the second point are always highlighted.
- *Releasing the mouse button* freezes the definition of the second point. The construction is then finished.

The figures below show the three stages during the construction of a line. Here the first point is a free point and the second point is the intersection of the two already existing lines.

Press the mouse ...   *... drag it ...*   *... and release*

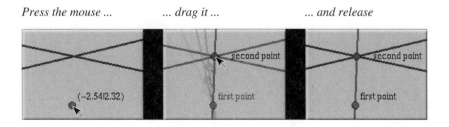

Synopsis

Create a line connecting two points by a press-drag-release sequence.

See also

- Add a point [p. 55]
- Line through point [p. 59]
- Add a parallel [p. 60]
- Add a perpendicular [p. 62]
- Add a circle [p. 64]
- Join [p. 76]

5.2.7.3 Line Through Point

This mode creates a line through a point with a certain slope. When the point is moved, then the slope of the line stays constant. However, in the move mode it is also possible to pick the line and change its slope. So this mode could also be called "Line by slope." The mode generates the line together with the point through which it passes by a single press-drag-release sequence. When the mouse is pressed the point is added. Dragging the mouse generates the line. It is always attached to the mouse pointer. When the mouse is released the position of the line is frozen, and the construction is finished. More exactly:

- *Pressing the left mouse button* generates the point. The definition of this point depends on the position of the mouse at the moment when the button is pressed:
    - If the mouse pointer is over an already existing point, then this point is taken.
    - If the mouse pointer is over the intersection of two elements (line, circle, or conic), then this intersection is automatically constructed and taken as the point.
    - If the mouse pointer is over just one element (line or circle), then a point is constructed that is constrained to this element. This point is taken as the point.
    - Otherwise a free point is added.
- *Dragging the mouse* generates the line. The slope can be adjusted freely. The line also snaps to already existing points, if they are selected by the current mouse position.
- *Releasing the mouse button* freezes the definition of the line. The construction is then finished. Depending on the final position of the mouse pointer two things can happen.
    - If the mouse pointer is over an already existing point, then this point is used as a second point on the line. The line is then the *join* (connecting line) of the first and the second point.
    - Otherwise a "Line with slope" is added.

Synopsis

Create a line through a point by a press-drag-release sequence.

See also

- Add a line [p. 57]
- Circle by radius [p. 65]

5.2.7.4 Add a Parallel

With this mode you can construct a line through a point which is parallel to another line with a press-drag-release sequence. The point through which the parallel should pass can also be generated in this mode. Constructing the parallel is a three step procedure.

- *Move the mouse over the line* for which you want a parallel. Press the left mouse button. This creates the parallel and the point through which it should pass.
- *Hold the left button and drag the mouse.* This moves the parallel and the new point to the desired position.
- *Release the mouse.* Now the construction is frozen. Depending on the position where you release the mouse the definition of the new point is adapted.
  - If the mouse pointer is over an already existing point, then this point is taken.
  - If the mouse pointer is over the intersection of two elements (line, circle, or conic), then this intersection is automatically constructed and taken as the new point.
  - If the mouse pointer is over just one element (line or circle), then a point is constructed that is constrained to this element. This point is taken as the new point.
  - Otherwise a free point is added.

The figures below show the three stages during the construction of a parallel. Here the new point will be bound to the already existing point *P*.

Press the mouse ...   ... drag it ...   ... and release

Synopsis

Create a parallel with a press-drag-release sequence.

Caution

The behavior of this mode is dependent on the geometry that is chosen. While in Euclidean Geometry there is always exactly one parallel, in non-euclidean geometries it is subject to the definition of parallelity. Depending on the underlying "philosophy", in hyperbolic geometry there can be infinitely many parallels, from an incidence geometric viewpoint, or there can be exactly two parallels, from an algebraic or measurement based point of view. *Cinderella* takes the algebraic point of view: *A parallel to a line **L** is a line that has a zero angle with **L***. So in hyperbolic geometry the mode produces exactly two parallels.

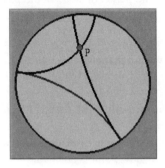

Hyperbolic Parallels

In elliptic geometry the usual viewpoint is that there are no parallels at all. However, from an algebraic standpoint there exist such parallels, they just have complex coordinates, in other words, they will never be visible. *Cinderella* constructs these parallels, and they are invisible but are nevertheless present in the "Construction Text" view [p. 100]. From there they can be accessed for further constructions.

See also

- Add a line [p. 57]
- Add a perpendicular [p. 62]
- Define Parallel [p. 77]

5.2.7.5 Add a Perpendicular

This mode allows a perpendicular to a line and through a point to be constructed with a press-drag-release sequence. The usage of the mode is exactly analogous to the "Add a parallel" mode. The point through which the perpendicular should pass can also be generated in this mode. Constructing the perpendicular is a three step procedure.

- *Move the mouse over the line* for which you want a perpendicular. Press the left mouse button. The perpendicular and the point through which it should pass are created.
- *Hold the left button and drag the mouse.* You can move the perpendicular and the new point to the desired position.
- *Release the mouse.* The construction is frozen. Depending on the position at which you release the mouse the definition of the new point is adapted.
    - If the mouse pointer is over an already existing point, then this point is taken.
    - If the mouse pointer is over the intersection of two elements (line, circle, or conic), then this intersection is automatically constructed and taken as the new point.
    - If the mouse pointer is over just one element (line or circle), then a point is constructed that is constrained to this element. This point is taken as the new point.
    - Otherwise a free point is added.

Synopsis

Create a perpendicular with a press-drag-release sequence.

Caution

The definition of perpendicularity depends on the chosen geometry.

See also

- Add a line [p. 57]
- Add a parallel [p. 60]
- Define Perpendicular [p. 78]

5.2.7.6 Add a Line With Fixed Angle

This mode allows a line that has a fixed, numerically given, angle to another line to be constructed. The new line and a point through which it should pass can be added with a press-drag-release sequence. The point through which the new line should pass can also be generated with this mode.

When you select this mode, a little window pops up asking you to specify the desired angle.

Input Window:

After that the usage of this mode is analogous to the "Add a parallel" mode.

- *Move the mouse over the line* to which you want the new line. Press the left mouse button. The new line and the point through which it should pass are created.
- *Hold the left button and drag the mouse.* You can move the new line and the new point to the desired position.
- *Release the mouse.* Now the construction is frozen. Depending on the position where you release the mouse the definition of the new point is adapted.

Synopsis

Create a line with fixed angle with respect to another line with a press-drag-release sequence.

Caution:

The definition of angle depends on the chosen geometry.

See also:

- Add a line [p. 57]
- Add a parallel [p. 60]

5.2.7.7 Add a Circle

This mode allows a circle, given by its center and a point on its perimeter, to be created. The mode generates the circle together with the two points with just one mouse click. When the mouse is pressed the center is added. Dragging the mouse generates the perimeter point and the circle. When the mouse is released the position of the perimeter point is frozen, and the construction is finished. The logic behind this mode is absolutely analogous to the "Add a line" modus.

Constructing a circle in this mode can be described as a three step procedure:

- *Pressing the left mouse button* generates the center point. The definition of this point depends on the position of the mouse at the moment when the button is pressed, like in the "Add a line" mode [p. 57].
- *Dragging the mouse* generates the circle and the perimeter point. Depending on the mouse position the definition of the perimeter point is chosen. As in the "Add point" mode the second point snaps to already existing elements. The choice of definitions is completely analogous to the choice for the first point. The elements that define the perimeter point are highlighted.
- *Releasing the mouse button* freezes the definition of the perimeter point. The construction is then finished.

Synopsis

Create a circle with a press-drag-release sequence.

Caution

The "shape" of a circle depends on the choice of the geometry.

See also

- Add a line [p. 57]
- Circle by radius [p. 65]
- Circle by fixed radius [p. 66]
- Circle by three points [p. 73]
- Compass [p. 71]

5.2.7.8 Circle by Radius

This mode allows a circle, given by its center and its radius, to be added. When the center of such a circle is moved, then the radius remains constant. However, in the move mode it is also possible to pick the boundary of the circle itself and change the radius.

The mode generates the circle together with its center with just one press-drag-release sequence. When the mouse is pressed the center is added. Dragging the mouse generates the circle. Its boundary is always attached to the mouse pointer. When the mouse is released the position of the circle is frozen, and the construction is finished. The logic behind this mode is analogous to the "Line through point" modus.

Constructing a circle in this mode can be described as a three step procedure:

- *Pressing the left mouse button* generates the center. The definition of this point depends on the position of the mouse at the moment when the button is pressed, like in the "Add a line" mode [p. 57].
- *Dragging the mouse* generates the circle. The radius can be adjusted freely. The circle also snaps to already existing points, if they are selected at the current mouse position.
- *Releasing the mouse button* freezes the definition of the circle. The construction is then finished. Depending on the final position of the mouse pointer two things can happen.
  - Either a "Circle by radius" is added,
  - or, if the mouse pointer is over an already existing point, then this point is used as a perimeter point for the circle. The circle is bound to that point.

Synopsis

Create a circle with free radius with a press-drag-release sequence.

Caution

The actual "shape" of a circle depends on the choice of the geometry.

See also

- Add a circle [p. 64]
- Circle by fixed radius [p. 66]
- Line through point [p. 59]

5.2.7.9 Circle by Fixed Radius

This mode allows a circle that has a fixed, numerically given, radius to be created. The new circle and its center are added with a single press-drag-release sequence.

When you select this mode, a little window pops up asking you to specify the desired radius, similar to the "Line with fixed angle" mode. After that the usage of this mode is analogous to the "Add a point" mode.

Synopsis

Create a circle with numerically given and fixed radius with a press-drag-release sequence.

Caution

The "shape" of a circle depends on the choice of the geometry.

See also

- Add a circle [p. 64]
- Circle by radius [p. 65]
- Add a line with fixed angle [p. 63]

5.2.7.10 Midpoint

In this mode the midpoint of two points can be constructed with a press-drag-release sequence. Analogous to the "Add a line" mode also the two points can be added.

- *Pressing the left mouse button*
  ... generates the first point. As in the other interactive modes the definition of this point depends on the position of the mouse at the moment when the button is pressed.
- *Dragging the mouse*
  ... generates the second point together with the midpoint. The second point also snaps to already existing points or intersections, if they are selected at the current mouse position.
- *Releasing the mouse button*
  ... freezes the actual definition. The construction is then finished.

Synopsis

Create two points and their midpoint using a press-drag-release sequence.

Caution

This innocent looking mode becomes interesting in non-euclidean geometries. While in Euclidean Geometry there is always a unique, finite, midpoint, in hyperbolic or elliptic geometry there are always two. Both points will be generated in this mode. When working in the Poincaré disc model of hyperbolic geometry, the second midpoint of two points in the disc always lies outside the disc. The presence of this point does not matter a lot, but there are circumstances in which one should be aware of this fact.

See also

- Add a line [p. 57]

5.2.8 Definition Modes

The *definition modes* use a simple "define by selection" mechanism to construct new elements. You choose a certain definition mode by clicking on its icon in the tool bar, then you are asked to select a certain number of elements in the view. After enough elements are selected the newly defined element is added to the construction. Although these modes are slightly inconvenient compared to the *interactive modes*, it is sometimes unavoidable to use these modes. There are four major circumstances when the definition modes should be used:

- A certain geometric operation is only provided as a definition mode. (This is the case for "Angular bisector", "Compass", "Circle by three points", "Center", " "Conic by five points", "Polar of a point", "Polar of a line", and "Polygon".
- A point of intersection could not be reachable in the visible area of a view. For instance, two lines can be almost parallel, so that their point of intersection lies far outside the viewing region in the window. Then the usual "Add a point" or "Add a line" modes are not applicable. But you can still select the lines and apply the "Meet" mode.
- An element could be invisible or complex. It can still be selected (e.g., in a "Construction Text" view) and a definition mode can be used.
- A situation could be ambiguous, for instance three lines passing through a point. Then the interactive "Add point" mode cannot be used to add a point of intersection (since in these cases all three lines would be close to the mouse pointer position). In this case use the "Meet" mode and select the lines whose intersection is desired.
  In the case of a *geometric theorem* where three lines always meet, such as the altitudes of a triangle, **Cinderella's** theorem checking mechanism ensures that the added point is considered incident to all three lines.

A few things are common to all definition modes and should be known in advance.

- Elements are selected by clicking them with the mouse pointer. The selection can be made in any of the geometric views, in particular in the *"Text view"* [p. 100] where all elements are always visible and selectable).
- Selected elements are highlighted in the view.
- Preselected elements are often considered when a definition mode is chosen. So if two lines are selected and you press the button that chooses the "Meet" mode then the point of intersection is instantly added.
- If you made a mistake in selecting an element you can deselect it with a second click.

- You can also select the elements by moving the mouse while holding the mouse button down. All elements that are touched by the mouse pointer will be selected (resp. deselected).
- Selected elements that do not contribute to the required selection of the mode are ignored. For instance, if you choose the "Circle by three points" [p. 73] mode, all selected lines, circles and conics are ignored.
- Selection modes communicate with the user by using the message line. There you can find messages that tell you about the next expected input.

5.2.8.1 Center

This mode constructs the center of a conic. In particular you can use this mode to construct the center of a circle. For general conics the center is the intersection of its axes of symmetry. The center of a conic is sometimes useful for constructing a point on a movable conic.

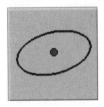　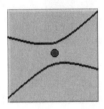

Center of an ellipse　　*Center of an hyperbola*

Synopsis

Select a conic and construct its center.

Caution

This mode is not properly supported in non-euclidean geometries, it always constructs the euclidean center.

5.2.8.2 Angular Bisector

This mode is used to construct the angular bisector of two lines. The application of this mode requires a little care, since two lines do not have only one angular bisector - *they have two!* To take this fact into account, this mode is provided with a position sensitive selection mechanism.

To define the angular bisector, two lines have to be selected. In order to indicate which angular bisector should be chosen three points are relevant: the click point of the first selection, the click point of the second selection, and the intersection of the two selected lines. Imagine a triangle formed by these three points. The inner angle at the intersection point of the lines will be bisected. This is what you would intuitively expect.

To make the selection process a bit simpler **Cinderella** gives graphical hints, as to which angle will be bisected.

First selection:
A line is highlighted.
The selection point
is memorized

Moving the mouse:
An indication of
the chosen angle
is given.

Second Selection:
The angular bisector
is added.

Synopsis

Select two lines and construct their angular bisector.

Caution

The definition of angular bisector depends on the type of geometry (Euclidean, hyperbolic or elliptic). In hyperbolic geometry angular bisectors can even have complex coordinates.

5.2.8.3 Compass

The compass is a very useful tool for transferring the distance between two points to some other place. *Cinderella's* compass works exactly like a real compass. You select a first point by clicking (i.e. you poke the needle into the first point). You select the second point by clicking (i.e. you adjust the compass to the distance between the first and the second point). Then you are ready to transfer the distance to another place. When you click at a third point a circle with the specified distance around this point is added to the construction.

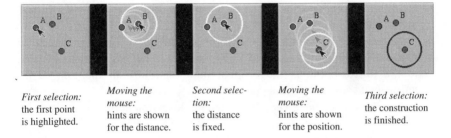

| First selection: the first point is highlighted. | Moving the mouse: hints are shown for the distance. | Second selection: the distance is fixed. | Moving the mouse: hints are shown for the position. | Third selection: the construction is finished. |

Synopsis

First select two points whose distance is the radius of a new circle centered at a third point.

Caution

The definition of a circle changes with the type of geometry (Euclidean, hyperbolic or elliptic).

See also

- Add a circle [p. 64]
- Circle by radius [p. 65]

5.2.8.4 Mirror

The mirror is a multi-purpose tool for doing reflections at points, lines or circles. The first mouse-click selects the "mirror". The following clicks select the elements that should be reflected. Reflected elements are either points, lines, or conics. You can deselect the mirror by clicking on it a second time. Depending on the choice of the mirror different actions are performed.

- *If the mirror is a line* then the usual reflection is taken. The mirror image of a point with respect to a line is a point that has the same distance to the line as the original point and lies on the perpendicular of the line that goes through the original point.
- *If the mirror is a point* then the reflection "at this point" is taken. The mirror image of a point with respect to a point is a point that has the same distance to the mirror-point as the original point and lies on the join of the mirror-point and the original point.
- *If the mirror is a Euclidean circle* then the inversion at that circle is taken. The inverse of a point with respect to a circle is a point that lies on the join of the original point and the center of the circle. The distance of the inverse point to the center of the circle is such that the product of this distance and the distance of the center to the original point is the circle's radius squared.

Reflections of lines or conics are considered as pointwise reflections.

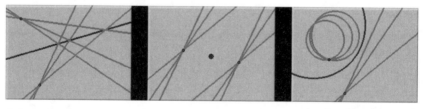

Reflection at a line   Reflection at a point   Inversion at a circle

Synopsis

First select a mirror. Then select elements that should be reflected.

Caution

The definition a mirror image heavily depends the choice of the underlying geometry.

5.2.8.5 Circle by Three Points

This mode is for constructing a circle that passes through three points. In Euclidean Geometry such a circle is always uniquely defined to be the circumcircle of the triangle defined by the three points. You choose the points one after another. The definition phase does not supply graphical hints.

Synopsis

Select three points. Then their circumcircle is constructed.

Caution

This mode is only available in Euclidean Geometry. In other geometries (hyperbolic, elliptic) there is no unique circle with this property.

See also

- Add a circle [p. 64]
- Conic by five points [p. 73]
- Circle by radius [p. 65]

5.2.8.6 Conic by Five Points

This mode constructs the unique conic that passes through five points. It is the basic mode for constructing a general conic.

Synopsis

Select five points to create a conic.

Caution

If four of the five points are collinear the conic is no longer unique and a null-element is computed.

See also

- Circle by three points [p. 73]

5.2.8.7 Polar of a Point

This mode constructs the polar of a point with respect to a conic. The point and the conic can be selected in any order. The polar, which is a line, is then constructed.

This mode has a few very interesting special cases that deserve some extra attention. If the point is on the conic itself then the unique *tangent* to the conic that passes through that point is constructed. Even more special, if the conic is a circle and the point lies on the circle then the *tangent* to the circle that passes through the point is constructed. Although the mode gives access to general polars this special case will most probably be its main application.

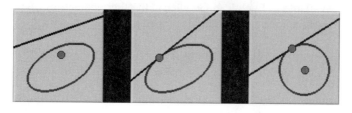

General Polar *Tangent to conic* *Tangent to circle*

Synopsis

Select a point and a conic and construct the polar line of the point.

See also

- Polar of a line [p. 74]

5.2.8.8 Polar of a Line

This mode constructs the polar of a line with respect to a point. The line and the conic can be selected in any order. This mode can be used to construct the point where a tangent line touches the conic.

Synopsis

Select a line and a conic to create the polar point of the line.

See also

- Polar of a point [p. 74]

5.2.8.9 Polygon

In this mode a polygon can be constructed from a sequence of vertices. Select the vertices in the order in which they should appear on the boundary of the polygon. Unlike the other definition modes, this mode does not expect a fixed number of input selections. You finish the creation of a polygon by selecting the first point of it a second time to close the polygon. For instance, if you want to construct a polygon through vertices *A*, *B*, *C* and *D*, you have to select *A*, *B*, *C*, *D* and *A* in this order.

If you made a mistake by selecting a wrong point, you just click the point again to delete it from the boundary of the polygon. However, only the last point in the sequence can be deselected in this way.

Graphical hints guide you through the definition phase. They always resemble the part of the polygon that has been constructed so far.

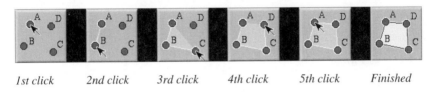

1st click   2nd click   3rd click   4th click   5th click   Finished

Synopsis

Select a sequence of points to define a polygon. To finish the definition select the first point again.

Caution

This mode is only supported in the Euclidean, spherical and textual view. The hyperbolic view ignores polygonal objects.

The orientation of the polygon is important when measuring its area.

See also

- Area [p. 82]

5.2.8.10 Join

In this mode the line joining two points can be created. Select two points and the line connecting them is constructed.

This mode may at first sight seem to be unnecessary. You usually add lines with the interactive "Add a line" [p. 57] mode. However, under certain circumstances it can be unavoidable to use the join mode. For instance, even if a point is not reachable (or even complex) in a usual view, it is still listed in the "Construction Text" view where it can be selected and used for the join mode.

Synopsis

Select two points and construct their join.

See also

- Add a line [p. 57]

5.2.8.11 Meet

In this mode the point of intersection of two line is constructed by selecting the lines.

This mode may at first sight seem to be unnecessary. Usually points are added using the "Add point" mode, or as a byproduct of some other interactive mode. However, under certain circumstances it can be unavoidable to use the "Meet" mode. For instance when the intersection of two lines is not reachable in a usual view, the lines are still listed in the "text view" [p. 100] where they can be selected and used for the "meet" mode. Another case where the meet mode is a good choice is in the case of ambiguities when three lines pass through one point. Then the meet mode is a secure way of adding a certain point of intersection.

Synopsis

Select two lines and create their intersection.

See also

- Add a point [p. 55]

5.2.8.12 Define a Parallel

This mode constructs the parallel of a line through a point. Select the point and the line in arbitrary order. In many cases you can use the more comfortable interactive "Add a parallel" mode.

Synopsis

Select a line and a point and construct the parallel to the line through the point.

Caution

The behavior of this mode is very dependent on the current geometry. Whereas in Euclidean Geometry there is always exactly one parallel, in non-euclidean geometries the number of parallels depends on the definition of parallelity. **Cinderella** takes the algebraic point of view: *A parallel to a line L is a line that has a zero angle with L.* Therefore the above mode produces exactly two parallels in hyperbolic geometry.

In elliptic geometry, under the usual definition, there are no parallels at all. However, from an algebraic standpoint there exist such parallels; they just have complex coordinates. In other words, they will never be visible. **Cinderella** constructs these parallels. They are invisible but still present in the "Text view" [p. 100], where they can be used for further constructions.

See also

- Add a parallel [p. 60]
- Define a perpendicular [p. 78]

5.2.8.13 Define a Perpendicular

This mode constructs the perpendicular of a line through a point. Select the point and the line in any order. In most cases you can use the more comfortable interactive "Add a perpendicular" mode.

Synopsis

Select a line and a point and construct the perpendicular to the line through the point.

Caution

The definition of perpendicularity depends on the chosen geometry.

See also

- Add a perpendicular [p. 62]
- Define a parallel [p. 77]

5.2.9 Measurements

Measuring is an important part, perhaps even the origin, of geometry. ***Cinderella*** has modes for measuring distances, angles and area. Their behavior, at least in Euclidean Geometry, is relatively straight forward:

- *Select two points and get their distance from each other.*
- *Select two lines and get the angle between them.*
- *Select a polygon, circle or conic and get its area.*

For most purposes that is all you have to know. However, as usual, there are a lot of fine details that sometime make things harder. If you deal with non-euclidean geometries their crucial defining property is a strange way to measure things. What goes on exactly is described in more detail in the *Behind the scenes* section. For now it is sufficient to know that in non-euclidean geometries measurements differ from the usual Euclidean measurements. Values of distances or angles may even become complex numbers. These values can be calculated with the theory of *Cayley-Klein-Geometries*. The treatment of measurements in such a general way is one of the core features of ***Cinderella***. You should not worry about the strange behavior of measurement in non-euclidean geometries. Perhaps the best way to understand it is to play with different constructions to get a feeling for the unique aspects. Allowing people to get an intuitive feel for non-euclidean geometries is one of the main goals of ***Cinderella***.

Another fine point comes from the measurements of areas. ***Cinderella*** is able to measure the area of polygons and of conics. In both cases there are a few things that deserve some extra attention. It is easy to define the area for a polygon, that does not overlap itself. But what happens if it does? The approach chosen in ***Cinderella*** is to use a general and consistent formula for area. Areas are counted with respect to an orientation. How much a point inside the polygon contributes to the area depends on its winding number with respect to the boundary.

The area of conics is also a delicate topic. It is easy to define the area of an ellipse. But what is the area of a hyperbola? Is it infinite? Is it undefined? Is it something completely different? ***Cinderella*** chooses an algebraic approach that tries to use only one formula for all different cases. It turns out, that the area of a hyperbola is most reasonably described by a complex number. So, if you make measurements of areas of conics do not be surprised if complex numbers sometimes show up.

Remark

The measurements are shown as texts. They can be used as "text objects", so they can be dragged around and be repositioned. Consult the description of the *add text* mode [p. 83] for these features.

80   Reference

5.2.9.1 Distance

Measuring distance in **Cinderella** is very much like drawing a line in the "Add a line" mode. You have to select the two points whose distance from each other you want to measure. This is done with a single mouse press-drag-release sequence.

- Move the mouse pointer over the first point and press it.
- You drag the mouse, with the left button pressed, to the second point. While you drag, a ruler with the actual distance from the first point to the mouse-pointer is shown.
- The ruler snaps to selected points. If the second point is selected, you can release the mouse button.

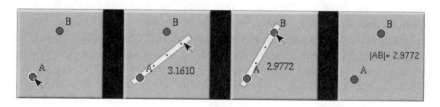

Mouse press: a point is selected.

Mouse drag: a ruler is shown.

Mouse drag: second point selected.

Mouse Released: Measurement added.

Synopsis

Measure the distance between two points with a press-drag-release sequence.

Caution

The definition of distance changes with the type of geometry (Euclidean, hyperbolic, or elliptic). In hyperbolic geometry distances may even be complex numbers.

5.2.9.2 Angle

This mode is used to measure the angle between two lines. The application of this mode requires a little care. Between two lines there is not only one angle - *there are two*, an angle and its counter-angle. To take this fact into account, the "Angle" mode is provided with a position sensitive selection mechanism. To measure the angle, two lines have to be selected. In order to know which angle will be chosen three points are relevant: the click point of the first selection, the click point of the second selection, and the intersection of the two selected lines. Imagine a triangle formed by these three points. The inner angle at the intersection point of the lines will be measured. This is what you would intuitively expect.

Cinderella simplifies the selection process with graphical hints which show the angle that will be measured.

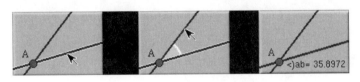

First selection:
a line is highlighted.
The selection point
is memorized.

Moving the mouse:
hints are shown
that indicate the
angle that will
be measured.

Second selection:
The angle is
measured.

Synopsis

Measure the angle between two lines.

Caution

The definition of angle depends on the type of geometry (Euclidean, hyperbolic, or elliptic). In hyperbolic geometry angles may even be complex numbers.

5.2.9.3 Area

In this mode you can measure areas of polygons and of conics. To measure the area of a polygon you simply click inside the polygon. To measure the area of a conic, you simply select it. In particular, you can use this mode to measure the area of a circle.

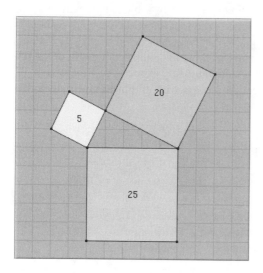

Pythagoras' Theorem with area measurements

Synopsis

Select a polygon or a conic and get its area.

Caution

Areas of hyperbola are complex numbers.

The contribution of a point to the area of a polygon is calculated with respect to its "winding" number. In particular areas may become negative if the polygon is oriented clockwise. If the polygon is self-overlapping areas may be zero.

5.2.10 Special Modes

5.2.10.1 ABC *Add Text*

The "Add text" mode is a multi-purpose mode for almost everything related to displaying text in a view. Here are the main applications of this mode:

- *Add plain text:* You click at some position in a view. Then an input dialog pops up asking you to input the desired text. When you exit the window, the text is displayed at the position where you clicked.
- *Edit text:* When you click on an already existing text you will get an editing window that shows the old text. This text can now be changed. After leaving the input window the old text is replaced by the new text.
- *Changing a label:* When you click on a geometric element you will also get an editing window, now showing the label of that element. You can now input a new label for the element. However, there is a restriction: The label of an element must be unique in the construction. If you try to input an already existing label a warning message is shown and the label remains unchanged.

Texts are more flexible objects than you may think at first. This comes from two important features, *docking* and *referencing*:

- *Docking:*
  Usually a text is located at a certain position relative to the coordinate system of the drawing. When you zoom or translate the view the text will follow the zooming. This is sometimes desired, but sometimes you might want a different effect. Imagine that you have a description text that should always be shown in the upper left corner of the view. Then you should use "docking." In the move mode you pick a text and move it around by dragging the mouse. If you come to a position close to the boundaries of the window you recognize that the text snaps to predefined docking positions. When the text is docked it stays in its position relative to the window.

  It is also possible to dock a text to a point. Drag the text close to the point until the point is highlighted. The text will then consistently stick to the point.
- *Referencing:*
  Often it is necessary to have changing parameters inside a text. Imagine a text saying: "The distance between A and B is 25 cm". In such a text you usually want to refer to the actual distance between "A" and "B" and not to a fixed string "25". You can do this by referencing the value of a geometric element. If "A" is the geometric element then "@#A" references its value. Moreover, you usually want to mark the fact that the "A" and the "B" in your text are labels. This forces further changes of element names are reported to

the text object, and that the displayed text is changed accordingly. You can use "@$A" to refer to a label. If "`dist`" is the label of the distance object then the above effect would be generated by the following input string:

```
The distance between @$A and @$B is @#dist cm.
```

In the case of points, lines and conics the "@#" operator refers to the coordinates. So, you can write:

```
The coordinates of @$A are @#A.
```

Altogether you have three possibilities for referring to the data of a geometric element. If `element` is the name of the label then you access

- the *label* of the element by `@$element`,
- the *defining algorithm* of the element by `@@element`, and
- the *value or position* of the element by `@#element`.

The texts that are then generated are precisely the ones you can see also in the Construction Text" [p. 100] view. The exact representation of the element's value or position can be influenced by the according settings in the "Format" menu.

If you want to include greek characters you can do so by using @ and the name of the character, like in `@alpha` or `@Omega`.

Synopsis

Add and edit a text.

5.2.10.2 Locus

A locus is the trace of a point under the movement of another point. A locus is defined by three objects:

- The *mover*, a free element whose movement drives the generation of the locus.
- The *road*, an element incident to the mover. The mover will be moved along the road.
- The *tracer*, the element whose trace is calculated and presented as a geometric locus.

The mover, road and tracer must be selected in this order. However, if the mover is either a "Point on line", a "Point on circle", or a "Line through point" then **Cinderella** automatically recognizes that there is a unique road and selects it for you automatically. You should watch the message line to check which input is required at a certain stage of a construction. Currently the following combinations of mover and road are supported:

- *mover = point, road = line:* The point moves along the line.
- *mover = point, road = circle:* The point moves along the circle.
- *mover = line, road = point:* The line rotates around the point.

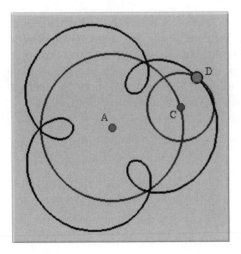

A cycloid

Cinderella also supports the selection of a "line" as the tracer. In this case the envelope of the moving line is calculated.

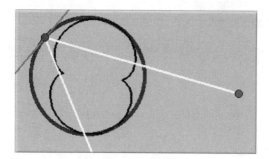

The envelope of light rays in a circular mirror

Loci of elements are real branches of algebraic curves. *Cinderella* will always try to generate an entire branch.

Synopsis

Generate a geometric locus by selecting a mover, a road and a tracer.

Caution

Sometimes you will notice a short delay when a locus is being calculated. Do not blame your computer or *Cinderella* (or, even worse, the authors). These delays can be caused by extremely difficult calculations that are necessary to get the correct result (a complete branch of the curve) or the correct screen representation after a movement. We tried our best to speed up these calculations, but there is a (mathematical) limit where we do not want to sacrifice accuracy for speed.

See also

- Animation [p. 87]

5.2.10.3 Animation

In an animation you do not move the points by yourself; *Cinderella* does it for you. An animation is defined by a "mover" and a "road", similar to the definition of a locus.

- The *mover* is a free element whose movement drives the animation.
- The *road* is an element incident to the mover. During the animation, the mover will be moved along the road.

Either you select the mover and the road in this order, or you click on a locus, which will select the mover and road of the locus. If the selected mover is either a "Point on line", a "Point on circle", or a "Line through point" then *Cinderella* automatically recognizes that there is a unique road and selects it for you. Currently the following combinations of mover and road are supported:

- *mover = point, road = line:* The point moves along the line.
- *mover = point, road = circle:* The point moves along the circle.
- *mover = line, road = point:* The line rotates around the point.

After you have started the animation an Animation Control panel pops up:

This panel contains 5 buttons and one slider. With the slider you can control the animation rate. The detailed meaning of the buttons is as follow:

- Start the animation.
- Pause the animation.
- Stop the animation.
- Export the animation to an HTML-page.
- Exit the animation.

When an animation is running all other functions of *Cinderella* are blocked. The only way to leave the animation is to press the "Exit" button.

In HTML-exports of animations which are generated by *Cinderella*, it is not possible to interactively drag the elements. All animation is done automatically by *Cinderella*. The export function exports all currently open views to an HTML page. The animation runs simultaneously in all these views. For further information consult the chapter on creating interactive web pages and exercises [p. 110].

Synopsis

Start an animation by selecting a mover and a road.

See also

- Locus [p. 85]

5.2.10.4 Add a Segment

This mode is completely analogous to the "Add a line" mode. By a press-drag-release sequence you add two points and a segment joining them. In addition, segments can also be supplied with arrows at one or at both ends. Unlike a line these segments are *not active* as geometric elements. This means you cannot use them for further geometric constructions (like generating intersections). The segments are purely graphical elements.

There are two reasons why you would use this mode. The first of these reasons is obvious: You want to create an arrow. The second reason is much more subtle. Perhaps you have a construction in which you want to show more than one segment supported by the same line. Since **Cinderella** does not allow elements to be added more than once there is no chance of having two identical lines. The usual line clipping mechanism of **Cinderella** is insufficient for this task. To bypass this problem you have the possibility of using segments. To avoid mathematical inconsistencies they are treated as graphical elements with no further geometric functionality.

Drawing an arrow

To draw an arrow you draw a segment, select it, and open the Arrow Editor by choosing the "Properties/Arrow Type" item in the menu.

The Arrow Editor

The Arrow Editor is analogous to the Appearance Editor. Its choices apply to the currently selected segments. You have the possibility of adding an arrow at either end of the segment. Four different types of arrow heads are provided. Two sliders allow you to control the size and the position of the arrow.

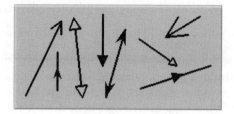

Examples of different arrows

Synopsis

Add a segment by a press-drag-release sequence.

Caution

Arrows are currently only supported in Euclidean views. Segments and arrows are *purely graphical* elements and they cannot be used for further constructions.

See also

- Add a line [p. 57]

5.3 Geometries

It is one of ***Cinderella's*** main features that it supports different kinds of *geometries*. If you are not accustomed to the idea that there are "different kinds of geometry" this may sound confusing. You should read the section *Behind the scenes* [p. 27] to get a feeling for the underlying ideas and how they are implemented in ***Cinderella***. Users who already have some basic knowledge about Euclidean and non-euclidean geometries can be satisfied reading this reference part only.

Another warning message: Do not confuse *geometries* and *views* [p. 94] - they both come in similar flavors, but the first defines the behavior of elements, while the second describes the presentation of elements.

5.3.1 Types of Geometries

In each main window of ***Cinderella*** you will find three buttons for choosing the type of geometry. In the present version Cinderella provides three different kinds of geometry: *Euclidean Geometry, Hyperbolic Geometry* and *Elliptic Geometry*. You can switch between these three geometries by pressing the buttons

- **Euc** for *Euclidean Geometry*.
- **Hyp** for *Hyperbolic Geometry*.
- **Ell** for *Elliptic Geometry*.

The choice of a new geometry does not affect the behavior of the elements you have already constructed. However, every newly added element is interpreted with respect to the new geometry. You can think of each element as having an entry that tells it to which geometry it belongs. The basic notions affected by the choice of the geometry are the measurements of *distances* and *angles*.

However, also other constructions are influenced by this choice. For instance, the "angular bisector" is defined to be a line whose angles to two other lines are equal. If the measurement of angles has changed, then the definition of "angular bisector" has to change, too. Similar things happen to "parallels" and "perpendicular lines". Also the definition of a circle is influenced by the geometry. A circle is the set of all points that have the same distance to the center. If the notion of "distance" is changed, the concept of "circle" changes as well.

Other operations are not affected by the choice of the geometry at all - the line joining two points will always be the same no matter in which of the above geometries you are.

The following list collects all constructions that are influenced by the choice of the geometry. Observe that the position, as well as the number, of elements that are constructed can change.

- *Distance:* The notion of distance depends on the geometry. It can even happen that in hyperbolic geometry distances of real points become complex numbers, for instance, when the line joining both points lies completely outside the horizon.
- *Angle:* The notion of angle depends on the geometry. As in the case of distances angles can also become complex numbers.
- *Circle:* The exact notion of circularity depends on the definition of "distance", which changes in each geometry. This influences all construction modes for circles. In the Euclidean view, hyperbolic or elliptic circles can look like arbitrary conics. The picture is clarified in the other views. In the hyperbolic view (Poincaré disc) hyperbolic circles really look like circles. In the spherical view elliptic circles look like circles on the surface of a ball.
- *Mirror:* The notion of reflection depends on distances and angles and thus it depends on the geometry. All kinds of mirrors are influenced this way.
- *Angular bisector:* In all three geometries, you have two angular bisectors for a line. However the exact position depends on the choice of the geometry. In hyperbolic geometry angular bisectors of real lines can become complex.
- *Midpoint:* The midpoint of two points depends on the definition of "distance". In euclidean geometry there is exactly one such midpoint. In hyperbolic and elliptic geometry there are two such midpoints (points of equal distance to the defining points). *Caution:* If you are in the "Hyperbolic view" only one of these points will be visible, since the other one lies outside the horizon.
- *Line with fixed angle:* This construction is influenced by the choice of the geometry because angles are involved.
- *Perpendicular:* The notion of a perpendicular depends on the notion of "angle" and is influenced by the choice of the geometry.
- *Parallel:* In **Cinderella** parallels of a line *L* are defined as lines that have an angle of zero to *L*. In Euclidean Geometry there is a unique parallel to *L* through a point. However, in hyperbolic and elliptic geometry there are two such parallels in general. The parallels in elliptic geometry will usually have complex coordinates, so that you will see them only in the "Construction Text" view.

In the present version of **Cinderella** some operations are not supported in all geometries. These operations are *Circle by three points*, *Area* and *Center*, where the Euclidean result is always calculated.

5.3.2 Views and Geometries

Although every geometry is usable together with every view a few combinations are a bit more common than others. Here is a short list of what these common choices represent.

- *Euclidean view [p. 94]* in *Euclidean geometry*:
  This may be the most common choice. The geometric elements behave as "usual elements" in a "usual plane".
- *Spherical view [p. 96]* in *Euclidean geometry*:
  This choice gives you control on the behavior "at infinity" of the Euclidean plane. The spherical view represents a double cover of the euclidean plane. Each line is mapped to a great circle and each point is mapped to an antipodal pair of points. The boundary of the non-rotated view corresponds to the "line at infinity" of the Euclidean plane.
- *Euclidean view [p. 94]* in *hyperbolic geometry*:
  What you see here is the so called "Beltrami-Klein" model of hyperbolic geometry. In these model hyperbolic lines are really straight. Measurement is done according to the definitions of the Cayley-Klein geometry. In the Euclidean view the horizon of hyperbolic geometry is shown as a thin circle.
- *Hyperbolic view [p. 98]* in *hyperbolic geometry*:
  This is what is known as the Poincaré disc. Hyperbolic lines are represented by circular arcs that cross the boundary of the disk at right angles. The Poincaré disc distorts the usual plane in a way that hyperbolic angles between lines correspond to "Euclidean" angles between the corresponding circular arcs. In mathematical terms: "The Poincaré disc is a conformal representation of the hyperbolic plane." In this picture hyperbolic circles really look circular.

  The whole disc represents only a part of the full plane of the corresponding Cayley-Klein geometry. The part that is shown corresponds to the region inside the circle shown in the Euclidean view.

  The measurement of distances is such that the distance from any interior point to any point on the boundary is equal to infinity.
- *Spherical view* in *elliptic geometry*:
  The spherical view is the natural view for elliptic geometry. The angle between two lines corresponds to the spherical angles of the corresponding great-circles. Measurement of distances corresponds to geodesic measurement of distances on the surface of a ball. Elliptic circles correspond to circles on the surface on the ball.

  However, one has to be a bit careful. Elliptic geometry is not equal to spherical geometry (geometry on the ball). This comes from the fact that in elliptic geometry antipodal points of the ball are identified with each other.

5.4 The Views

The entry "Views" in the menu bar offers items to open windows containing different views of the geometric construction. Each view is a kind of "projection" of the abstract configuration to some visible part of the computer screen. Usually you can make constructions and manipulations in any of the views. The changes are reported to the other views immediately. In particular, you can have many views of the same type (for instance two Euclidean views with different scales).

The views are also related to the different kinds of geometries. Which view is appropriate for which geometry is discussed in the section "Views and geometries".

5.4.1 Euclidean View

The Euclidean view is the usual drawing surface. When *Cinderella* is started you will get a window containing a Euclidean view. It is the natural window for doing Euclidean geometry.

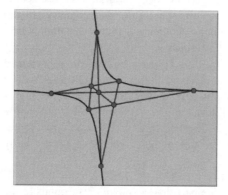

Pascal's Theorem in a Euclidean view

The Euclidean view has view-specific control buttons. They can be used for zooming and translation. Furthermore they are used to control grids and snap points.

5.4.1.1 Translate

This mode allows the entire coordinate system to be translated After you have selected this mode you can move the view around while pressing the left mouse button.

5.4.1.2 Zoom in

With this mode you can zoom into the drawing. There are two ways of doing this:

- With a press-drag-release sequence you mark a region. The view will be zoomed to show exactly this region.
- You can click the left mouse button over any position in the view. The view will be zoomed around the click point. The factor of zooming is 1.4. You can also click the right button (or shift-click the left button) to get the inverse of this zooming operation.

5.4.1.3 Zoom out

With this mode you can zoom out the drawing. The operation is inverse to the zoom in operation. There are two ways of using this mode:

- With a press-drag-release sequence you can mark a region in the view. The view will be appropriately zoomed so that the presently visible part zooms to that region.
- You can click the left mouse button at any place in the view. The view will be zoomed around the click point. The factor of zooming is 0.7. You can also click the right button (or shift-click the left button) to get the inverse of this zooming operation.

5.4.1.4 View All Points

Pressing this button adjusts the current zoom settings in order to show all points of the construction.

5.4.1.5 Toggle Grid

Shows/hides a grid on the view.

5.4.1.6 Toggle Axes

Shows/hides a coordinate system within the view.

5.4.1.7 Toggle Snap

Toggles the "snap" mode. In this mode the grid points are magnetic and will attract the mouse. This is the ideal tool for exact drawings. If you first select this mode the grid and axes are automatically shown. You can hide them again individually.

5.4.1.8 Denser Grid

Gives the grid a higher density.

5.4.1.9 Coarser Grid

Gives the grid a lower density.

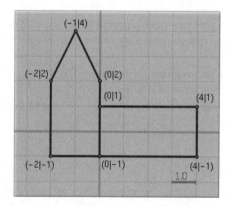

Using "Snap" for exact drawings

5.4.2 Spherical View

The spherical view arises from a projection of the Euclidean plane onto the surface of a ball. The center of the projection is the center of the sphere. The plane does not pass through this center.

5.4.2 Spherical View

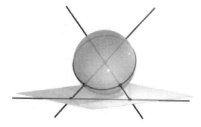

Projection from the plane onto the ball

This projection maps each point to an antipodal pair of points on the sphere. Each line is mapped to a great circle (an equator) on the sphere. The incidence structure is preserved. Working with the spherical view allows elements at infinity to be manipulated. They lie on the boundary of the image of the unrotated sphere.

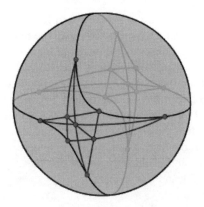

Pascal's Theorem in a spherical view

The spherical view is a natural way to represent elliptic geometry. Measurement of angles between lines corresponds to measuring angles on the sphere. Measuring distances corresponds to the usual geodesic measurement of distances on the sphere, while keeping in mind that antipodal points are identified with each other.

5.4.2.1 Rotate

This mode allows the spherical ball to be rotated (after the projection). In this mode you can rotate the view by moving the mouse keeping the left button pressed.

5.4.2.2 Spherical Reset

This action resets the rotation of the spherical view to its original position. After a reset the visible boundary of the ball corresponds to the line at infinity again.

5.4.2.3 The Scale Slider

This slider lets you control the distance from the sphere to the Euclidean plane. You can use this slider to find the right magnification of the drawing.

5.4.3 Hyperbolic View

The hyperbolic view is the natural view for hyperbolic geometry. In fact doing hyperbolic geometry is the main reason for opening a hyperbolic view. The hyperbolic view represents an implementation of the Poincaré disc model of hyperbolic geometry. In this model the (finite part of the) hyperbolic plane is represented by a disc. Each line is represented by a circular arc that is orthogonal to the boundary of this disc. The measurement of angles between lines is conformal. This means that you can read off angles by measuring Euclidean angles between the circular arcs. The measurement of distances is such that the elements on the boundary are "infinitely far away" from any other point on the disc. If you "walk" in hyperbolic unit steps in one direction, you will never reach the boundary. In the disc the steps seem to become smaller and smaller (in Euclidean measurement).

5.4.4 Polar Euclidean and Spherical View

Polarity is an important concept of projective geometry. Because of the complete symmetry between lines and points it is possible to turn every statement about incidences of points and lines into a corresponding "polar statement" where the role of points and lines are interchanged.

5.4.4 Polar Euclidean and Spherical View

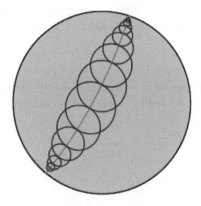
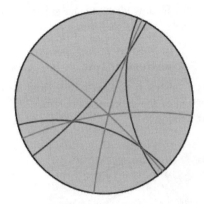

Hyperbolic circles of equal size   *Hyperbolic altitudes meet in a point*

Cinderella offers two polar views for visualizing polarity. The polarity implemented in **Cinderella** is the polarity with respect to the identity matrix. In algebraic terms, we use the homogeneous coordinates of a point and interpret them as a line, and vice versa. Geometrically this finds its easiest interpretation in the spherical view. Whenever you have a point consider it as a "north pole"; the corresponding "equator" is its polar line. Whenever you have a line consider it as an "equator"; the corresponding "north pole" is its polar point. The figure below shows a configuration in the spherical view and in the polar spherical view.

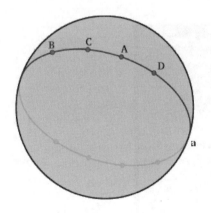
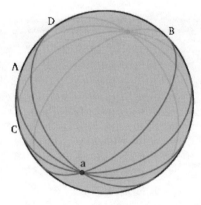

A configuration ...   *... and its polar*

Polar view elements are selectable but moving them is disabled. If you want to move elements then you have to control them in a primal view.

5.4.5 Construction Text

The Construction Text is a textual description of the construction steps. Each element of the geometric construction is represented by a row in the construction text window.

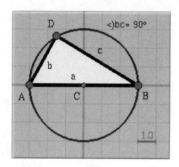

A picture of Thales' Theorem

Each row shows a little icon that resembles the element it refers to. The icon is shown in the size, color and shape of the element. This makes it easy to identify elements quickly.

Who?	What?	Where?
A	Point(-3|0)	(-3|0)
B	Point(3|0)	(3|0)
a	Join(A,B)	y = 0x + 0
C	Mid(A,B)	(0|0)
C0	Circle(C,B)	$(x+0)^2 + (y+0)^2 = 3^2$
D	PointOn(C0,-60.16°)	(-1.49|2.60)
b	Join(A,D)	y = 1.73x + 5.18
c	Join(D,B)	y = -0.58x + 1.74
Poly0	Polygon(D,A,B)	—
alpha0	Angle(b,c,D)	90°
Text0	Text(D,alpha0)	<)bc= 90°

Move the view by dragging the mouse

The construction steps for Thales' Theorem

The construction text consists of four columns. The first column shows the icons. The second column shows the labels of the geometric elements. These labels are unique identifiers within the construction. The third row explains how the element is defined. The fourth row represents the current value of the element. In most cases this is the present location with respect to the coordinate system. With the menu *"Format"* you can change how the locations or values of elements are shown.

The texts that appear in the "Construction Text" are exactly the three entries that can be referenced by the "Add text" mode [p. 83].

The four columns are separated by vertical lines. You can pick these lines with the mouse to control the width of each column. If the text has too many rows for the window, a scrollbar at the right side of the window is available for scrolling.

5.4.6 General Functions

In each view a tool bar at the bottom displays view-specific operations. The ones that are common in all views are described below.

5.4.6.1 *Generate PostScript*

If you press this button the contents of the view will be exported to a PostScript file. You will be asked whether you want to have the picture in color, gray or black and white. The file generated contains an initial section where you can adjust the appearance of the printout later. It starts with:

```
%%%%%%%%%%%%%%%%%%%%%%%%%%%%%%%%%%%%%%%%
% Drawing mode:                         %
% mode=0: color  / mode=1: gray  / mode=2: bw %
%%%%%%%%%%%%%%%%%%%%%%%%%%%%%%%%%%%%%%%%
/mode 0 def
```

By changing the mode flag you can later on choose the color representation you want. It is also possible to change the background color.

```
%%%%%%%%%%%%%%%%%%%%%%%%%%%%%%%%%%%%%%%%
% Background color  (default white)     %
%%%%%%%%%%%%%%%%%%%%%%%%%%%%%%%%%%%%%%%%
/background { 1 1 1} def
```

The other settings in the PostScript file are more or less self-explanatory.

5.4.6.2 [Euc] [Hyp] [Ell] *Choose the Geometry*

With these buttons you change the current geometry. All geometric constructions refer to this geometry. For a detailed discussion of geometries consult the chapters "Geometries" [p. 91] and "Behind the scenes" [p. 27].

5.5 The Appearance Editor

The Appearance Editor is a window in which you can control the graphical appearance of elements in the construction. Changes always refer to the selected elements. Any changes made in the appearance are instantly passed to the views.

The settings in the appearance panel also serve a second purpose. They are taken as default settings for any newly added elements. This means newly added elements get the appearance that corresponds to the settings in the Appearance Editor.

The attributes that you can control with the appearance editor are the following:

- color of the geometric elements [p. 103],
- default colors of the views (background, highlight and text) [p. 105],
- line clipping [p. 105],
- labeling of elements [p. 107],
- pinning (is an element movable or not?) [p. 107],
- the amount of overhang of a line [p. 108],
- size [p. 108],
- and opaqueness [p. 109].

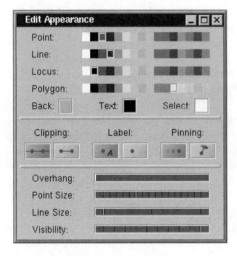

The Appearance Editor Window.

5.5.1 Color

The colors of the geometric elements are organized as follows. The geometric elements are grouped into four different types:

- points
- lines (including conics and circles)
- loci
- polygons

For each of these groups the Appearance Editor offers a palette of 16 different default colors.

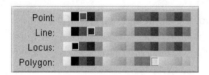

Color palettes for the elements.

In each palette one entry is marked. This is the current default for this group. You can change this mark by clicking on another entry. So, if you want to change the color of a point from red to green, you have to do the following:

- Select the point.
- Open the appearance editor (if it is not already open).
- Click the green box in the palette called "Point".

You can also change the appearance of several points at the same time if you select more than one point. After your change has been performed the green color is the new default for added points. A similar procedure applies to the other three color groups.

It may be the case that you are not satisfied with the color choices offered by *Cinderella*. You can change a color value by double-clicking on its entry in the color palette. A color chooser window pops up in which you can adjust the red/green/blue values of the palette entry.

Color Chooser

It is important to know that by changing the color of an entry in the palette *you change the color of all geometric elements associated with this color entry.*
Currently the palette is not stored with the configuration in the ".cdy" file.

5.5.2 View Colors

There are three other entries in the Appearance Editor that control colors.

View Colors

These entries refer to the default colors for the views.

- *Back:* This is the background color of the views.
- *Text:* This is the color used for the texts and labels. This color is also used for the boundary of points.
- *Select:* This is the color used for highlighting elements.

These settings apply to all views. Make sure that the three colors are clearly distinguishable from each other.

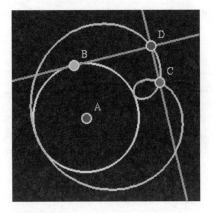

Example of an individualized color map

5.5.3 Clipping

You can control whether a line is "clipped" or not. For this purpose the Appearance Editor offers two buttons that turn clipping on or off.

Clipping choice

A non-clipped line will be drawn over the whole size of a view. A clipped line is truncated with respect to the points lying on it. The rules that govern the clipping logic are a bit intricate, but they lead to natural behavior:

- To be a "clipping point" of a line the point must be incident to the line. Furthermore the point must not be completely invisible and it must not lie at infinity. Hidden clipping points can be created by setting the size of the point to zero [p. ??].
- All clipping points of a line are consulted for the clipping.
- If the line has at least two clipping points then the portion of the line that reaches from the first clipping point of the line to the last clipping point is drawn.
- If there are less than two clipping points the line is not clipped.

The above rules have the effect that

- at least a small portion of the line is always shown,
- and that all visible points incident to the line lie on this portion.

This is what you would expect for geometric drawings. The decision of whether a point is (always) incident to a line is done by the automatic theorem checking of *Cinderella*. In this way a correct and mathematically consistent behavior is ensured.

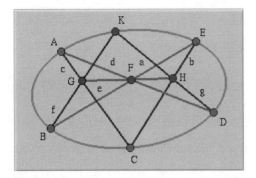

Pascal's Theorem with clipping

5.5.4 Labelling

Every geometric element has a flag to indicate whether its label should be shown or not. This can be controlled by the Label choice:

Label choice

So far the display of labels in geometric views is only provided for points and for lines. If you want to label conics or loci you can place a point of size zero near them. This point serves as a carrier for the label.

5.5.5 Pinning

Pinning describes the behavior, not the appearance of an element. If a movable element (e.g., a free point) is "pinned" then it is no longer movable.

Pinning choice

This feature is necessary to restrict the freedom of an element in a drawing and can be useful when you design exercises.

The following types of elements are movable and can therefore be pinned.

108   Reference

- free points
- points on a line
- points on a circle
- lines with a slope
- circles by radius

5.5.6 Overhang

It is often not desirable to have a clipped line end directly at some points. It looks much nicer to have some overhang that suggests that the line continues further. You can control the size of an overhang by using the overhang slider. The slider's position adjusts the overhang on both sides between 0% and 50% of the line's total length.

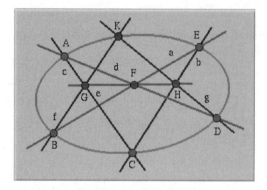

Pascal's Theorem with clipping and overhang

5.5.7 Size

To control the size of objects there are two sliders in the Appearance Editor. One slider controls the size of point elements.

Possible point sizes

Points of size zero are very useful objects. They are invisible, but they are selectable and movable and they still serve as active clipping points for clipping lines. This makes them ideal elements to use when you want to have a line that you can move by grabbing its endpoints. So you do not have to make unnecessary "decorations" with visible points.

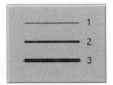

Possible line sizes

The other slider controls the thickness of line-like elements (between 1 and 3). "Line-like elements" are lines, circles, conics and loci.

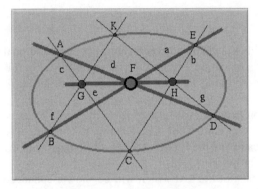

Pascal's Theorem with points and lines of different sizes.

5.5.8 Opaqueness

The Appearance Editor has a slider for adjusting the opaqueness of elements. The opaqueness can vary from "absolutely invisible" to "solid".

Grades of opaqueness.

Absolutely invisible elements are good for auxiliary constructions that should not disturb the rest of the drawing. Slightly visible elements can be used for parts of a construction that are of minor importance. For an element to take active part in the views (being movable and selectable) it needs a opaqueness of at least 20 percent. You can select invisible elements in the textual view. They cannot, however, be used for clipping lines.

6 Creating Interactive Webpages and Exercises

The most exciting feature of **Cinderella** is its ability to create interactive webpages. You can publish any construction, even those using several views, within seconds, and without further knowledge about HTML.

This chapter explains the three scenarios for export: Plain examples, animations and construction exercises. You also find detailed information on the exported HTML code and instructions for post-processing of the web pages, e.g. adding explanatory text.

6.1 Glossary

If you are familiar with the World Wide Web and its technical background or if you do not want to bother with technicalities right now you can safely skip this section. As an aid for the further description we want to explain a few of the terms used below.

HTML is the page description language (or format) for web pages. It can be created and edited with any plain text editor, but it is much more convenient to use a special HTML editor. You can view the HTML code of a web page with the "view source" option in your web browser.

The HTML code mainly consist of the text which will be shown, enhanced by *tags* which describe the appearance and structure of this text. Here is an example:

```
<p>This is a paragraph containing some <b>bold text</b>
and some <i>italics text</i>.</p>
<p>This picture <img src="pappos.gif"> was produced
  with <a href="http://www.cinderella.de">Cinderella</a>!</p>
```

This fragment describes two paragraphs marked by `<p>...</p>`. The first paragraph contains two regions which should be typeset with special fonts, the first one, marked by `<b>...</b>`, in bold, the second one in italics. You might recognize the easy structure of HTML, most of the *elements* are marked with opening (`<something>`) and closing tags (`</something>`).

The second paragraph in the example shows how to include an image using the `<img>` tag. This tag does not have a closing companion, but it uses an option (`src=...`) to reference the image file. You also find a *hyperlink*, which is the most powerful element in HTML. You name a location which can be reached by clicking on the phrase included by `<a>...</a>`.

Hyperlinks are usually given by an *URL*, an uniform resource locator. These describe resources by the protocols which give access to them. For the WWW most communication is done with the *Hypertext Transfer Protocol*, short for http. This explains the http:-part of the WWW addresses, which can be left out since most browsers assume it as a default. Other examples are ftp, the *File Transfer Protocol*, or the prefix file: which describes files that reside locally on your harddisk.

A special tag is reserved for Java integration into web pages. Whenever a Java compatible browser encounters an `<applet>` tag it tries to load the java program referenced by the `code` option and runs it inside a rectangle on the web page. The size of the rectangle is given by the `width` and `height` options. These programs are called *Applets*, as opposed to standalone applications. The diminutive is a little misleading, since applets can be as powerful as full applications.

The applet tag can also contain an `archive` option which describes the location of the java code. For *Cinderella* we provide an archive called `cindyrun.jar` which contains all the code needed for showing and manipulating constructions. This is an example how the applet tag produced by *Cinderella's* web export functions could look like:

```
<applet code    = "de.cinderella.CindyApplet"
        archive = "cindyrun.jar"
        width   = 435
        height  = 231>
<param...
</applet>
```

You find many `<param>` tags which pass additional parameters like the filename of the construction to the applet. Never change or delete these parameters without exactly knowing what they mean.

6.2 Exporting Plain Examples

This is the easiest way to create a webpage with *Cinderella*. Whenever you have created a configuration you can use the export button to create an interactive webpage showing this construction in exactly the views that you are currently using. Each view will be a separate applet, and these applets communicate using a kernel ID which you find as a parameter of the applet.

The construction itself is not saved in the HTML code, but in a separate file with the extension ".cdy". Whenever you create a web page you are prompted to save your file, if you have not done so before. The applet expects the file in the same directory as the html file of the web page.

Next you will be asked for a filename of the web page. This should end in ".html" or ".htm", depending on your local standards. If you do not supply one of these extensions, *Cinderella* will assume ".html" as a default. This step finishes the web export, and you should be able to view the result in a Java-1.1 compatible web browser, after you have copied the runtime library into the export directory.

More specifically: The applet expects a file called "cindyrun.jar" in the same directory, which contains the necessary code to show and manipulate constructions. You find this file in the installation directory and you have to copy it into the directory containing the interactive web page. Consult the section on advanced options for further help.

If you experience any problems, be sure to check whether:

1. The file ending in ".htm" or ".html" exists and is readable.
2. The file mentioned as value in <param name=filename> section of the html file exists and is readable (you should be able to load it with *Cinderella*)
3. The file "cindyrun.jar" is present in the same directory as the two files above and is referenced in the archive option of the applet tag.
4. You are using a Java-1.1 compatible browser. We recommend using Netscape Version 4.08 or 4.51 or Internet Explorer 4.0 or higher, but due to the rapid changes in the browser industry you should look up the list of recommended browsers at the *Cinderella* homepage (www.cinderella.de). Your installation CD contains a recent version of Netscape, but you should be aware that neither Springer-Verlag nor the *Cinderella*-authors can support this third-party program. Please visit the Netscape home page for information on their browser. On MacOS you should use Internet Explorer, since Netscape does not support Java 1.1 on MacOS (March 1999).

The exported construction is always in move mode. That means that movable elements can be dragged around within the applet rectangle. If you want to prevent elements from being movable, please use the pinning option in the Appearance Editor [p. 103].

6.3 Exporting Animations

Exporting automatic animations is similar to and as easy as exporting interactive examples. Start the desired animation using the animation mode [p. 87] and adjust its speed using the slider. Then use the export button in the animation control dialog to export. All rules for exporting interactive examples still apply: You need three files for successful loading of the animation in a browser, the web page (.html), the construction (.cdy) and the runtime library (cindyrun.jar).

While exporting animations please keep in mind that the potential visitors of your web page might have slower computers than you. You should adjust the animation speed accordingly.

6.4 Creating Interactive Exercises

You can use *Cinderella* to create interactive exercises for students. However, the design of a good exercise is much more involved than the creation of an interactive construction. The export itself is easy, but for a educationally valuable exercise you should have some experience with geometry, *Cinderella* and teaching.

For a complete exercise three major steps are needed: Constructing, defining the input, solution(s) and hint(s), and the actual export.

6.4.1 Exercise Construction

You cannot create an exercise without solving it yourself in the beginning. While this is just fair, it has a more important technical reason. *Cinderella* uses your construction for checking the correctness of the students' solutions.

So, before you can define an exercise, you have to do an example construction. In this chapter we will use an easy example, construction of the midpoint: For two points *A* and *B*, find the point in the middle between them, using only ruler (lines) and compass (circles). Here is how you can solve this task:

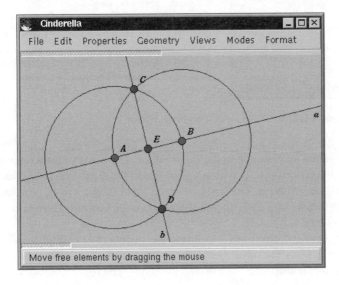

6.4.2 Editing the Exercise

6.4.2.1 Defining the Input

After you did the example construction open the exercise design dialog via the appropriate button in the toolbar 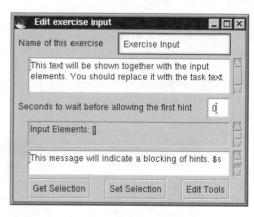 or the entry in the file menu. Next, open the Input Editor by double-clicking the "Exercise Input" entry. Now you will have three windows on your desktop.

The Exercise Editor   The Input Editor

    Change the texts shown in the Input Editor. The name of the exercise will be used as a reference only, it will not show up in the final exercise. Change it to "Midpoint", nevertheless. The first text in the Input Editor will be presented to the students as the exercise task, so you should enter the task text there (after deleting the default text).

    Later we will define hints which lead the students to a correct solution. If you want to block hints for some time after presenting the taks, enter the number of seconds in the appropriate field. You should also change the message in the last field, which will be shown when a hint is requested before the blocking time has passed. The string $s will be replaced by the number of seconds after which a new hint will be available.

    After these changes, the Input Editor should look approximately like this:

6.4.2 Editing the Exercise

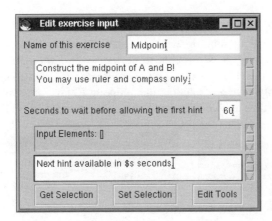

So far you have not defined the start elements for the construction. Since the students should start with the points *A* and *B* you select these points in the main construction window. Use the selection mode . Then you define the input with the "Get Selection" button in the Input Editor. "A" and "B" show up as input elements.

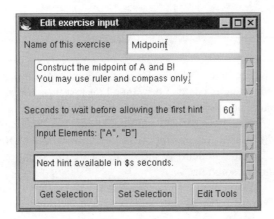

Now you have defined everything you need as input for the exercise. The hints and the solution are still missing. If you want, you can close the Input Editor now.

6.4.2.2 Defining Solutions

Next we will define the solution, which will later be used for checking the correctness of the exercises. It is possible to define several alternative solutions for ambiguous exercises, like "define an angular bisector of lines l and g".

Here we only have one solution, the midpoint E. First press "Add Sol." in the Exercise Editor, then double-click on "Solution #1" to open a Solution Editor.

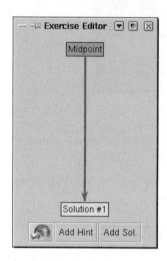
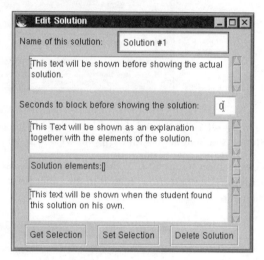

The first text area in the Solution Editor can contain a text that will be shown when a hint is requested. If you do not want a hint text then you should delete the text contained there, otherwise you can replace it with a custom message. We will replace the text after we have defined additional hints. Until then, just erase it.

The second text area contains the message that will be shown when the solution is presented to the student. It should describe the situation that you will present as an example solution.

The last text area is used for a message that will be shown when the student completes the exercise. You should replace it with a friendly and encouraging message!

We still have to define the solution elements. As we said above this is only E. Select this point and click on the "Get Selection" button. It should appear as a solution element in the editor window.

6.4.2 Editing the Exercise

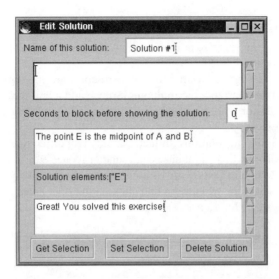

If there were several solutions for this exercise we could add another solution, which would be accepted by the automatic theorem checking engine like the first one. When the students requests hints, only the first solution will be used. An example for an exercise with multiple solutions is "Construct a equilateral triangle over the segment AB"; there is a triangle above and a triangle below the segment.

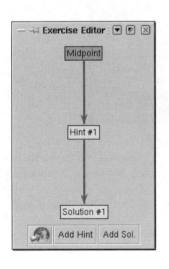
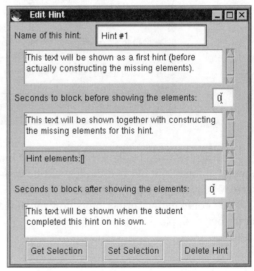

6.4.2.3 Defining Hints

You could export the exercise at this moment. However, we want to define some hints for the students, too.

An important step in the midpoint construction are the two circles. Their radical axis will pass through the midpoint. So, we add a hint for this. Add a hint in the Exercise Editor with the "Add Hint" button and double-click the "Hint #1" to open a Hint Editor.

Hints are similar to solutions. In fact, the only real difference is that the exercise is not yet solved when a hint has been found or requested.

Replace the default texts with your custom messages, like you did with the solution. Then, select the two circles and click on the "Get Selection" button. The screenshot shows the Hint Editor after these actions.

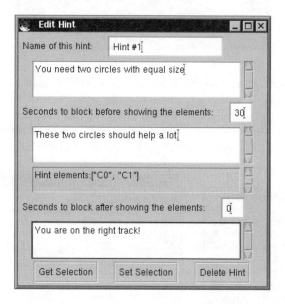

The first text will be shown when the first hint is requested. When the student requests another hint the two circles will be constructed, but this will work only when 30 seconds have passed since the first hint.

You could define further hints now by clicking the "Add Hint" button again, but for our example this should be sufficient.

6.4.2.4 Tool Selection

You did not define the available tools for the exercise yet. In that case a default toolbox will be used, which includes the "Add point", "Add a line", "Compass" and "Move" modes, and action buttons for "Undo", "Give Hint" and "Restart Exercise". If you want to add other tools or remove some of the above you can do so by clicking the "Edit Tools" button in the Input Editor.

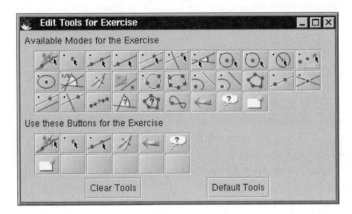

You add a tool by clicking on the appropriate button in the upper half of the Tool Editor, you remove it by clicking on its buttons in the lower half, which displays the tools that will be available.

For the example you might want to replace the compass by the "Add Circle" mode, but you can also accept the tools as they are.

6.4.3 Saving and Creating the HTML

When you are satisfied with the exercise you can create the corresponding HTML code for the web page with the export button .

The export is similar to the plain example or animation export. You have to save the construction and you have to save the HTML code into the same directory. *Cinderella* expects the `cindyrun.jar` file there, too.

Before the export the software will do some sanity checks on your exercise, for example it checks whether there are input elements and solution(s).

In addition to the currently open ports there will be two extra applets, an exercise console where the messages will be shown, and a control panel with the tools. The sizes of these applets are always the same, but you can edit them later directly in the HTML code, if you wish. Please refer to the tips and tricks section below.

6.4.4 Testing the Exercise

After you have exported the exercise you can visit the HTML file with your browser in order to test the exercise. You should ensure that your hints and solutions make sense.

If you want to change texts or elements you can load the .cdy-file of the exercise and use the Exercise Editor. All exercise data will be saved in the .cdy file and can be manipulated later.

6.4.5 Design Considerations

1. While **Cinderella** is good at proving it does a bad job on guessing. Your solutions should be well-defined: They should only rely on the start elements of the exercise. We implemented a "guessing" heuristic that tries to handle free helper elements, but this heuristic can fail. Example: If you use a point P on circle CO in your solution then **Cinderella** detects points on that circle and decides that the first of these must be P. But if you use two points on CO in a different way, then **Cinderella** has no chance to know the right assignment of two points in the student's solution to your points, and it makes an arbitrary decision. In most cases this heuristic will work just fine, but you should keep in mind that this might be the reason for unexpected results.
2. Write precise task texts, that clearly define the solution you expect. Define all possible solutions.
3. Mark the absolutely necessary elements only. In the midpoint example above, point E is sufficient to describe the solution. If you mark lines a and b too, since you thought that they are needed, you destroyed the chance of having another path to the solution. The screenshot below shows you how a student could have solved the exercise.

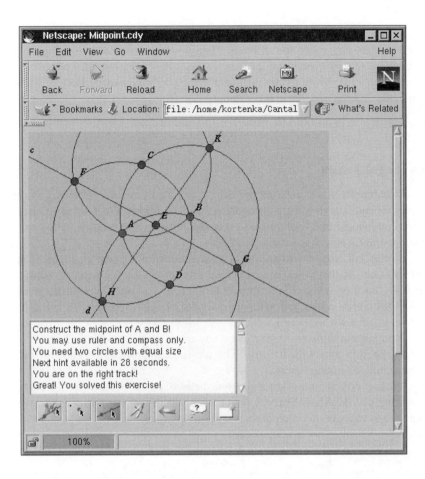

6.5 Post-Processing

The web site containing your construction or exercise is very basic. *Cinderella* does not try to be a full-featured web editor. You can use any other web editor for post-processing the HTML files.

The width and height parameters of the applets can be changed, if you want to. This is important for the exercise console and control panel, since their sizes are fixed within *Cinderella*. The view sizes, however, should be set to the correct size before you export the construction, since this is much easier.

Never change the kernelID parameter of the applet, it is important for inter-applet communication. The order and placement of the applets can be changed arbitrarily. You can also merge two different HTML pages, which gives you the possibility of showing different constructions on one web page.

If you have several *Cinderella*-enabled pages on your web site you can use a single `cindyrun.jar` for all these. Then you will have to change the archive parameter of all your applets to reflect the location of your central `cindyrun.jar`. It should be sufficient to include the complete URL of its location in the archive tag, like in this example:

```
<applet ... archive="http://www.yourweb.edu/directory/cindyrun.jar" ...>
```

6.6 Legal Issues

With *Cinderella* you purchased the license to redistribute the necessary runtime library for interactive web pages under certain conditions. We want to summarize these conditions, but keep in mind that the only legally binding terms can be found in the License Agreement [p. 126].

First of all, you should not try to make money out of your examples, that is, sell them or put them onto a commercial online service. You can certainly use them for teaching (even if you get paid for this). If you want to publish a book or CD-ROM which uses *Cinderella*, you should contact Springer-Verlag and the authors to get a written permission.

You are not allowed to give away any other files than `cindyrun.jar` of *Cinderella*, this covers the `cindy.jar` file (note the difference!) and the online documentation, among others.

Whenever you are unsure about your license, please contact Springer-Verlag in writing or via eMail. You can find contact details on the Springer website at http://www.springer.de or http://www.springer-ny.com.

7 Installation

This chapter should help you to install *Cinderella* on your computer.

7.1 General Information

You need two things to run *Cinderella*: the program code and a Java Virtual Machine (JVM). We included a installation software on the CD which includes JVMs for the most popular platforms. If your platform is not included, you can still be able to run *Cinderella*, provided that you get a Java-1.1 compatible JVM first. Please refer to the Javasoft website at http://www.javasoft.com for information on third-party Java ports and proceed to the "Java install" section in this chapter.

For you convenience we included some Java-1.1-compatible Internet browsers on the CD. This includes versions of Netscape Communicator for Unix and Windows, Microsoft Internet Explorer for MacOS and Microsoft Windows 95/98/2000/NT, as well as a preview version of iCab for MacOS. We do not offer support for these browsers.

7.2 Installing on Windows 95, 98 or NT 4.0

Insert the CD into the CD drive and open it. You will find a file called "install" in the root directory of the CD. Start this program and follow the instructions.

After you installed *Cinderella* you will be able to double-click files with the ".cdy" extension in order to view or edit them.

If you want to uninstall *Cinderella* you can do so by choosing the "Uninstall Cinderella" shortcut.

7.3 Installation on Unix Platforms

7.3.1 Installing on Sun Solaris (SPARC)

Mount the CD (e.g. using "volcheck", or ask your system administrator) and change to the Solaris directory, where you find a file called "install.bin". Run it with "./install.bin" or "sh install.bin".

Follow the on-screen installation instructions. You can choose the installation directory and link locations during the install process. Be sure to choose locations where you have write access.

If you already have a Java-1.1 compatible JVM you can choose to use it instead of the packaged one, which will reduce the disk space needed by *Cinderella*.

After you installed *Cinderella* you can run it by executing the "Cinderella" link or the file "Cinderella" in the installation directory.

For uninstalling the software you can use the link "Uninstall Cinderella" or the executable found in the UninstallerData subdirectory of the installation directory.

7.3.2 Other Unix-like Platforms

First of all you have make sure that a Java-1.1 JVM is available and properly installed on your system. Consult the Javasoft webpages for information on third-party Java ports.

Mount the CD and change to the Unix subdirectory on the CD (you should ask your System Administrator if you do not know how to mount a CD). Run the "install.bin" script you find there.

If the installation software finds more than one JVM it will ask you to choose one. After that, the installation procedure begins and works like the Solaris version. Please read the preceding section for more details.

7.3.3 Installing a JVM on Linux

In order to avoid unnecessary downloads we put versions of the Java Runtime Environment for the Linux operating system on the CD. Please understand that neither Springer-Verlag nor the authors can support this software, which is provided "as-is" and without any warranties.

Please consult the README file in the Linux subdirectory on the CD for more information, or consult the Java-Linux site at http://java.blackdown.org.

After you have installed a JVM you can proceed as written in the previous section, "Installing on Unix-like platforms".

7.4 Installing on MacOS

Insert the CD into your CD drive and open the CD icon that appears on your desktop. You will find three folders, called Cinderella, MRJ Install, and Internet Browsers. If you are unsure about your version of Macintosh Runtime for Java (MRJ), please change to the MRJ Install folder and follow the instructions given there.

If you have MRJ 2.2 or higher installed, you can run *Cinderella* directly from the CD by double-clicking the Cinderella-icon in the Cinderella folder. If you prefer, you can copy the entire Cinderella folder to your harddisk and run it from there. No other installation procedure is needed.

7.5 Installing on Other Java Platforms

It should be possible to install *Cinderella* on any platform that is capable of running Java-1.1. You will understand that we could not test the software on all these platforms, and your mileage might vary. We tried not to rely on any platform specific features which coulb be unavailable somewhere.

Insert the CD into your CD drive and find the directory called "Java". The file "install.zip" which is contained in this directory contains the complete installation software. Use your platform specific way to add install.zip to the Java classpath and run the class called "install". After that you can proceed as in the Unix install.

A possible way do the above would be to enter

```
jre -cp install.zip install
```

in a command shell. Please consult your JVM documentation for details.

7.6 Installing Using a Web Browser

If you are unsure about your platform or if you do not feel like using a command shell you can use a web browser for installation. Open the page "install.htm" in the "Webinstall" with a browser directory on the CD and follow the instructions.

7.7 Troubleshooting

If the installation fails for some reason, please check the following facts:

- The CD is in the drive and accessible (on Unix systems you have to mount it).
- You have enough space on your harddisk (at least 2 MB without Documentation, at least 7 MB with Documentation, if you install without a VM; installing a VM can use up more than 10 MB in addition).
- You have write permissions in the installation and link location.
- If you use a third-party JVM, make sure that it is Java-1.1 compatible. If possible, use one of the JVMs found on the CD.
- *Cinderella* is not yet installed or properly uninstalled.

If none of the above causes your trouble, please visit our website for a list of problems and solutions or mail to `install@cinderella.de` for help.

8 License Agreement

8.1 Conditions of Use and Terms of Warranty

This is an unauthorized translation of the original text in German language [p. 130]. Only the original text is legally binding.

§1 Concluding the Contract

Opening the sealed plastic cover binds the end user to the following conditions of use and the terms of the warranty. If the end user does not wish to be bound by these conditions, he should return the unopened package to his supplier or to Springer-Verlag and the selling price will be refunded. For the return of goods, §7 is valid.

§2 Copyright and Conditions of Use

1. All rights pertaining to the Software (program and handbook) are owned exclusively by Springer-Verlag. The Software is protected by copyright. Independent of this, the parties agree that the laws of copyright shall apply to the Software.
2. Springer-Verlag grants the end user, subject to legal liability, the non-exclusive right to use the Software as described by the terms of this contract. Under this contract use of the program is restricted to that carried out according to the instructions described in the handbook. Decompiling, disassembling, reverse engineering or in any way changing the program is expressly forbidden.
3. The program may, at any one time, only be used on one computer at a single workplace. When used on computers with several or many terminals or in a network, a license application must be made for each workstation or terminal on which use is possible.
4. The end user may not demand the handing over of the source program or information pertaining to it or the production documentation or the Software even when Springer-Verlag ceases to improve or update the Software.
5. The runtime libraries for interactive web pages ("cindyrun.jar") as mentioned in the manual may be redistributed unchanged together with interactive constructions, animations and exercises created by the Software, if
    1. no fee is charged, and
    2. a link to the web pages of the Software (http://www.cinderella.de) or equivalent pages of Springer-Verlag is provided on the page of the construction, animation or exercise, or on the page of contents referring to it.

This explicitly excludes the redistribution within commercial online services, CD-ROMs or as an add-on for books (even if there is no additional fee).

§3 Transfer of the Software

1. Any transfer (e.g. sale) of the Software to a third party and with it the transfer of the right and the possibility of its use may only occur with the written permission of Springer-Verlag.
2. Springer-Verlag will give this permission when the end user up to this point makes a written application and the subsequent end user makes a declaration that he will remain bound by the terms of this contract. Receipt of permission terminates the right of the first end user to operate the program and the transfer to the second end user may take place.

§4 Unauthorized Use

1. The complete Software is protected by the laws of copyright, the laws governing the use of trademarks, the laws of trade and commerce and this contract. Violations may lead to action being taken under civil and criminal law.
2. The buyer is liable to Springer-Verlag for any damages or detriment accruing from any infringement of these regulations.
3. If the customer violates the obligations set out in §§ 2 and 3, Springer-Verlag can withdraw the authority to use the program immediately without giving notice.

§5 Functional Limitations of the Software

1. Even with the latest state of technological development and with meticulous care being taken during production, errors in the Software cannot be excluded.
2. The Software tries to simulate geometrical constructions via a mathematical model. This approach suffers from numerical effects and the results must not be used without careful checking and deliberation at the responsibility of the user. The constructions done with the Software serve no other use than for teaching and there is no warranty - neither expressed nor implied - for correct calculations.
3. The hardware and basis Software described in the handbook are necessary for the functional capability of the program. The installation of the Software must be carried out exactly as described in the instructions in the handbook. Deviation from these instructions can lead to damage of the hardware and also to other software and data.
4. The runtime library for interactive web pages has been tested for best compatibility with current (November 1998) browsers. Nevertheless, neither

Springer-Verlag nor the authors can guarantee for the representation, correctness and usability of constructions and exercises in interactive WWW pages.

§6 Warranty

1. In response to justified claims, Springer-Verlag has, as first possibility, the option of supplying the user with another copy of the program (including another program release). If the claim is still not remedied, the end user can demand the return of the selling price from his supplier when he returns the Software in compliance to the terms set out in §7.
2. A prerequisite to making a claim under the warranty is that the end user supplies an exact description of the defect in writing.
3. The end user has no claim to a reduction in the selling price or to correction of defects. In other respects the German Code of Civil Law (BGB) concerning the warranty of goods shall apply (§§ 459 to 480 BGB).

§7 Returning the Software

1. The customer can only return the Software (e.g. according to §1 or 6 Sect. 1) in its entirety (in particular - with the handbook and the program discs) together with the original sales receipt/invoice. In addition he has to hand over the declaration in the handbook stating that no copies remain in his possession.
2. Should this declaration prove to be false, Springer-Verlag can demand a contract penalty of three times the recommended retail price and further compensation if necessary.

§8 Advice

1. Springer-Verlag has inaugurated the possibility of asking the author questions with reference to the Software. However, this is a voluntary service and is not the customer's right.
2. The questions can be concerned with installation, operation, and problems of utilization. Information on mathematical questions will not be given.
3. Questions should be mailed or sent via e-Mail to Springer-Verlag. The answers from the author are merely forwarded by Springer-Verlag without being checked. The questions are normally answered in the order they are received. It will not be possible to answer every question.

§9 Liability

1. Springer-Verlag and the author are only liable for willful intent, gross negligence, and when the program fails to fulfill its assured purpose and function. The assured purpose and functions are those which are explicitly declared in writing. There is no liability for information described in §8.

2. The liability under German law for product liability is unaffected. The plea that the end user is also at fault remains an option for Springer-Verlag.

§10 Conclusion

1. The location of the competent court for all legal action in connection with the Software and this contract is D-69121 Heidelberg if the contract partner is a registered trader or equivalent, or if he has no legal domicile in Germany.
2. This contract is exclusively governed by the law of the Federal Republic of Germany with the exception of the UNCITRAL laws of trade and commerce.
3. Should any provision of the contract prove unenforceable or if the contract is incomplete, the remaining provisions will remain unaffected. The invalid provision shall be deemed replaced by the provision which in a legally binding matter comes nearest in its meaning and purpose to the unenforceable provision. This shall apply to any omission in the contract that may occur.

8.2 Nutzungs- und Garantiebedingungen

§1 Vertragsabschluss

Durch Öffnen der Verpackung vereinbart der Endnutzer mit dem Springer-Verlag die nachfolgenden Nutzungs- und Garantiebedingungen. Falls der Endnutzer dies nicht anerkennen will, kann er die ungeöffnete Packung mit dem Original-Kaufbeleg binnen zwei Wochen gegen volle Erstattung des Kaufpreises seinem Lieferanten oder dem Springer-Verlag zurückgeben. Für die Rückgabe gilt §7.

§2 Urheber- und Nutzungsrechte

1. Alle Nutzungsrechte an der Software (Programme und Handbuch) stehen ausschließlich dem Springer-Verlag zu. Die Software ist urheberrechtlich geschützt. Unabhängig davon vereinbaren die Parteien hiermit, dass auf die Software die Regeln des Urheberrechts anzuwenden sind.
2. Der Springer-Verlag überlässt dem Endnutzer die nicht ausschließliche schuldrechtliche Befugnis, die Software vertragsgemäß zu nutzen. Vertragsgemäß ist nur eine Nutzung, bei der das Programm mit Hilfe der im Handbuch beschriebenen Anweisungen ausgeführt wird. Insbesondere sind das Verändern, Bearbeiten, Umgestalten und Dekompilieren der Software unzulässig.
3. Das Programm darf zur selben Zeit nur auf einem Rechner und auf einem Arbeitsplatz benutzt werden. Bei Nutzung auf Rechnern mit mehreren Arbeitsplätzen oder in Netzen muss pro Arbeitsplatz, auf dem die Nutzung möglich ist, eine Lizenz erworben werden.
4. Der Endnutzer kann die Aushändigung oder Kenntnisnahme des Quellprogramms oder der Herstelldokumentation der Software nicht verlangen, auch nicht, wenn der Springer-Verlag die Pflege der Software aufgibt.
5. Die im Handbuch genannten Laufzeitbibliotheken für interaktive WWW-Seiten ("cindyrun.jar") dürfen in unveränderter Form zusammen mit durch die Software erstellten interaktiven Konstruktionen, Animationen und Übungsaufgaben weitergegeben werden, sofern
    1. dafür keine Gebühren erhoben werden, und
    2. ein Verweis auf die WWW-Seiten der Software (http://www.cinderella.de) oder entsprechenden Seiten des Springer-Verlages auf der Seite der Konstruktion, Animation oder Aufgabe oder in einem übergeordneten Inhaltsverzeichnis gemacht wird.

   Dies schließt die Weitergabe innerhalb von kostenpflichtigen Online-Diensten und auf kommerziellen CD-ROMs sowie als Zusatz zu Büchern (selbst wenn keine spezielle Zusatzgebühr erhoben wird) ausdrücklich aus.

§3 Weitergabe der Software

1. Jede Weitergabe (z.B. Verkauf) der Software an Dritte und damit jede Übertragung der Nutzungsbefugnis und -möglichkeit bedarf der schriftlichen Erlaubnis des Springer-Verlages.
2. Der Springer-Verlag wird die Erlaubnis geben, wenn der bisherige Endnutzer dies schriftlich beantragt und eine Erklärung des nachfolgenden Endnutzers vorliegt, dass dieser sich an die Regelungen dieses Vertrages gebunden hält. Ab dem Zugang der Erlaubnis erlischt das Nutzungsrecht des bisherigen Nutzers und wird die Weitergabe zulässig.

§4 Unerlaubte Nutzung

1. Die gesamte Software ist durch Urheberrecht, Warenzeichenrecht, Wettbewerbsrecht und diesen Vertrag geschützt. Verstöße hiergegen können zivilrechtlich und strafrechtlich verfolgt werden.
2. Der Käufer haftet dem Springer-Verlag für alle Schäden und Nachteile aufgrund von Verletzungen dieser Regelung.
3. Wenn der Kunde gegen die Pflichten der §§ 2 und 3 verstößt, kann der Springer-Verlag die Nutzungsbefugnis aus wichtigem Grund fristlos kündigen.

§5 Funktionsbeschränkungen der Software

1. Nach dem Stand der Technik können Fehler der Software auch bei sorgfältiger Erstellung nicht ausgeschlossen werden.
2. Die Software versucht mit einem mathematischen Modell geometrische Zusammenhänge realistisch darzustellen. Die dabei gewonnene Darstellung unterliegt numerischen Effekten und darf daher nur nach vorheriger eigenverantwortlicher Prüfung weiterverwendet werden. Insbesondere sind die mit der Software erstellten Konstruktionen nur für Lehrzwecke geeignet und es wird keine Garantie für die korrekte Berechnung übernommen.
3. Für die Funktionsfähigkeit des Programms ist die im Installationshandbuch beschriebene Hardware und Basissoftware notwendig. Die Installation der Software muss genau nach den Vorschriften des Handbuchs erfolgen. Abweichungen hiervon können zu Schäden an der Hardware, an anderer Software und an Daten führen.
4. Die Laufzeitbibliotheken für interaktive Webseiten wurden auf eine größtmögliche Kompatibilität mit derzeitigen (November 1998) Webbrowsern getestet. Dennoch übernimmt weder der Springer-Verlag noch der Autor eine Garantie für die Darstellung, Korrektheit und Benutzbarkeit von Konstruktionen und Aufgaben in interaktiven WWW-Seiten.

§6 Garantie

1. Bei berechtigten Beanstandungen hat der Springer-Verlag zunächst die Möglichkeit, dem Endnutzer ein anderes Exemplar zu überlassen (auch ein anderes Programm-Release). Wenn damit die Beanstandung nicht behoben ist, kann der Endnutzer den Kaufpreis zurückverlangen, wenn er die Software entsprechend §7 zurückgibt.
2. Die Inanspruchnahme der Garantie setzt voraus, dass der Endnutzer den Mangel schriftlich genau beschreibt.
3. Auf Minderung und Nachbesserung hat der Endnutzer keinen Anspruch. Im übrigen gelten die Regeln der kaufmännischen Gewährleistung (§§ 459-480 BGB) entsprechend.

§7 Rückgabe

1. Der Kunde kann die Software (z.B. nach §1 oder 6 Abs. 1) nur komplett (insbesondere mit Handbuch und Programmdisketten) und mit dem Original-Kaufbeleg zurückgeben. Zusätzlich hat er zu erklären, dass keine Kopien existieren.
2. Für den Fall der Unrichtigkeit dieser Erklärung kann der Springer-Verlag eine Vertragsstrafe in Höhe des dreifachen empfohlenen Richtpreises der Software und gegebenenfalls weiteren Schadenersatz verlangen.

§8 Beratung

1. Der Springer-Verlag eröffnet die Möglichkeit, Fragen in bezug auf die Software an den Autor zu richten. Ein Rechtsanspruch für diesen Dienst besteht jedoch nicht.
2. Die Fragen können die Installation, die Handhabungs- und Benutzungsprobleme des Programms betreffen. Auskünfte über mathematische Fragestellungen werden nicht erteilt.
3. Anfragen sind schriftlich oder über e-Mail an den Springer-Verlag zu richten. Der Springer-Verlag vermittelt lediglich ungeprüft die Beantwortung durch den Autor. Die Antworten erfolgen üblicherweise in der Reihenfolge des Eingangs. Nicht jede Frage wird beantwortet werden können.

§9 Haftung

1. Der Springer-Verlag und der Autor haften nur bei Vorsatz, bei grober Fahrlässigkeit und bei Eigenschaftszusicherungen. Die Zusicherung von Eigenschaften bedarf der ausdrücklichen schriftlichen Erklärung. Für Auskünfte nach §8 wird nicht gehaftet.
2. Die Haftung aus dem Produkthaftungsgesetz bleibt unberührt. Der Einwand des Mitverschuldens des Endnutzers bleibt dem Springer-Verlag offen.

§10 Schluss

1. Gerichtsstand für alle Klagen im Zusammenhang mit der Software und dieser Vereinbarung ist Heidelberg, wenn der Vertragspartner Vollkaufmann oder gleichgestellt ist oder keinen allgemeinen Gerichtsstand in Deutschland hat.
2. Es gilt ausschließlich das Recht der Bundesrepublik Deutschland mit Ausnahme der UNCITRAL-Kaufgesetze.
3. Sollte eine Bestimmung dieses Vertrages unwirksam sein oder werden oder sollte der Vertrag unvollständig sein, so wird der Vertrag im übrigen Inhalt nicht berührt. Die unwirksame Bestimmung gilt als durch eine solche Bestimmung ersetzt, welche dem Sinn in Zweck der unwirksamen Bestimmung in rechtswirksamer Weise wirtschaftlich am nächsten kommt. Gleiches gilt für etwaige Vertragslücken.

8.3 Java(tm) Runtime Environment

This is the Java Runtime Environment license that applies if you installed a JVM together with *Cinderella*.

                    Java(tm) Runtime Environment

                         Version 1.1.7B

                       Binary Code License

This binary code license ("License") contains rights and
restrictions associated with use of the accompanying Java
Runtime Environment Version 1.1.7B software and documentation
("Software"). Read the License carefully before using the
Software. By using the Software you agree to the terms and
conditions of this License.

1.   License to Distribute. Licensee is granted a
royalty-free right to reproduce and distribute the Software
provided that Licensee: (i)distributes the Software complete
and unmodified, only as part of, and for the sole purpose of
running, Licensee's Java applet or application ("Program")
into which the Software is incorporated; (ii) does not
distribute additional software intended to replace any
component(s) of the Software; (iii) does not remove or alter
any proprietary legends or notices contained in the
Software; (iv) only distributes the Program subject to a
license agreement that protects Sun's interests consistent
with the terms contained herein; (v) may not create, or
authorize its licensees to create additional classes,
interfaces, or subpackages that are contained in the "java"
or "sun" packages or similar as specified by Sun in any
class file naming convention; and (vi) agree to indemnify,
hold harmless, and defend Sun and its licensors from and
against any claims or lawsuits, including attorneys' fees,
that arise or result from the use or distribution of the
Program.

2.   Restrictions. (a) Software is confidential copyrighted
information of Sun and title to all copies is retained by
Sun and/or its licensors. Except as otherwise provided by
law for purposes of decompilation of the Software, Licensee
shall not translate, reverse engineer, disassemble,
decompile, or otherwise attempt to derive the source code of
Software. Software may not be leased, assigned, or
sublicensed, in whole or in part, except as specifically

authorized in Section 1. (b) Software is not designed or
intended, and Sun expressly disclaims any representations or
warranties (either expressed or implied), for use (i)in
online control of aircraft, air traffic, aircraft navigation
or aircraft communications; or (ii) in the design,
construction, operation or maintenance of any nuclear
facility.

3. Trademarks and Logos. This License does not authorize
Licensee to use any Sun name, trademark or logo. Licensee
acknowledges as between it and Sun that Sun owns the Java
trademark and all Java-related trademarks, logos and icons
including the Coffee Cup and Duke ("Java Marks") and agrees
to comply with the Java Trademark Guidelines at
http://java.sun.com/ trademarks.html.

4. Disclaimer of Warranty. Software is provided "AS IS,"
without a warranty of any kind. ALL EXPRESS OR IMPLIED
REPRESENTATIONS AND WARRANTIES, INCLUDING ANY IMPLIED
WARRANTY OF MERCHANTABILITY, FITNESS FOR A PARTICULAR
PURPOSE OR NON-INFRINGEMENT, ARE HEREBY EXCLUDED.

5. Limitation of Liability. IN NO EVENT WILL SUN OR ITS
LICENSORS BE LIABLE FOR ANY LOST REVENUE, PROFIT OR DATA, OR
FOR DIRECT, INDIRECT, SPECIAL, CONSEQUENTIAL, INCIDENTAL OR
PUNITIVE DAMAGES, HOWEVER CAUSED AND REGARDLESS OF THE
THEORY OF LIABILITY, RELATING TO THE USE, DOWNLOAD,
DISTRIBUTION OF OR INABILITY TO USE SOFTWARE, EVEN IF SUN
HAS BEEN ADVISED OF THE POSSIBILITY OF SUCH DAMAGES.

6. Termination. Licensee may terminate this License at any
time by destroying all copies of Software. This License will
terminate immediately without notice from Sun if Licensee
fails to comply with any provision of this License. Upon
such termination, Licensee must destroy all copies of
Software.

7. Maintenance and Support. No upgrades or support are to
be provided to Licensee under the terms of this License.

8. Export Regulations. Software, including technical data,
is subject to U.S. export control laws, including the U.S.
Export Administration Act and its associated regulations,
and may be subject to export or import regulations in other
countries. Licensee agrees to comply strictly with all such
regulations and acknowledges that it has the responsibility
to obtain licenses to export, re-export, or import Software.
Software may not be downloaded, or otherwise exported or
re-exported (i) into, or to a national or resident of, Cuba,

Iraq, Iran, North Korea, Libya, Sudan, Syria or any country to which the U.S. has embargoed goods; or (ii) to anyone on the U.S. Treasury Department's list of Specially Designated Nations or the U.S. Commerce Department's Table of Denial Orders.

9. Restricted Rights. Use, duplication or disclosure by the United States government is subject to the restrictions as set forth in the Rights in Technical Data and Computer Software Clauses in DFARS 252.227-7013(c) (1) (ii) and FAR 52.227-19(c) (2) as applicable.

10. Governing Law. Any action related to this License will be governed by California law and controlling U.S. federal law. No choice of law rules of any jurisdiction will apply.

11. Severability. If any of the above provisions are held to be in violation of applicable law, void, or unenforceable in any jurisdiction, then such provisions are herewith waived or amended to the extent necessary for the License to be otherwise enforceable in such jurisdiction. However, if in Sun's opinion deletion or amendment of any provisions of the License by operation of this paragraph unreasonably compromises the rights or increase the liabilities of Sun.

9 References

[Bel1]

E.T. Bell, *"The Development of Mathematics,"* Dover Publishing, New York, 1992 (orig. 1945).

[Bel2]

E.T. Bell, *"Men of Mathematics,"* Touchstone Books, New York, 1986 (orig. 1945).

[CrRG]

H. Crapo, J. Richter-Gebert, *"Automatic proving of geometric theorems,"* in *"Invariant Methods in Discrete and Computational Geometry,"* Neil White ed., Kluwer Academic Publishers, (1995).

[Cox1]

H.S.M. Coxeter, *"Projective Geometry," (2nd ed.)* Springer, New York, Berlin, 1994 (orig. 1963).

[Cox2]

H.S.M. Coxeter, *"The Real Projective Plane," (3rd ed.)* Springer, New York, Berlin, 1992 (orig. 1949).

[Gre]

M.J. Greenberg, *"Euclidean and non-Euclidean Geometries," (3rd ed.)* Freeman and Company, New York, 1996 (orig. 1974).

[Kl1]

F. Klein, *"Development of Mathematics in the 19th Century,"* Math. Sci. Press, 1979 (orig. 1928).

[Kl2]

F. Klein, *"Vorlesungen über nicht-euklidische Geometrie,"* Springer, Berlin, reprinted 1968 (orig. 1928).

[Kor]

U. Kortenkamp, *"Foundations of Dynamic Geometry,"* Ph.D. thesis, ETH Zürich 1999.

[KRG]

U. Kortenkamp, J. Richter-Gebert, *"Foundations of Dynamic Geometry,"* in preparation.

[Lab]

J.M. Laborde, *"Exploring non-euclidean geometry in a dynamic geometry environment like Cabri-Géomètre,"* in "Geometry Turned On," J. King, D. Schattschneider (eds.), Math. Assoc. of America, 1997, pp. 185-191.

[RG]

J. Richter-Gebert, *"Mechanical theorem proving in projective geometry,"* Annals of Mathematics and Artificial Intelligence, *13*, 1995, pp. 139-172.

[Stru]

D.J. Struik, D.L.Struik, *"A Concise History of Mathematics,"* Dover Publishing, 1987.

[Yag]

I.M. Yaglom, *"Felix Klein and Sophus Lie - Evolution of the Idea of Symmetry in the Nineteents Century,"* Birkhäuser, Boston, Basel, 1988.